# Introduction to Nanophotonics

# Introduction to Nanophotonics

Editor:

Juan Liu

Intelliz Press

www.intellizpress.com

Published in 2022 by
Intelliz Press LLC
1 Radisson Plaza # 800
New Rochelle, New York
NY 10801
United States of America

**Introduction to Nanophotonics**
**Editor: Juan Liu**

ISBN: 978-1-68251-819-9



For more information about Intelliz Press and its products, visit our website at **www.intellizpress.com**

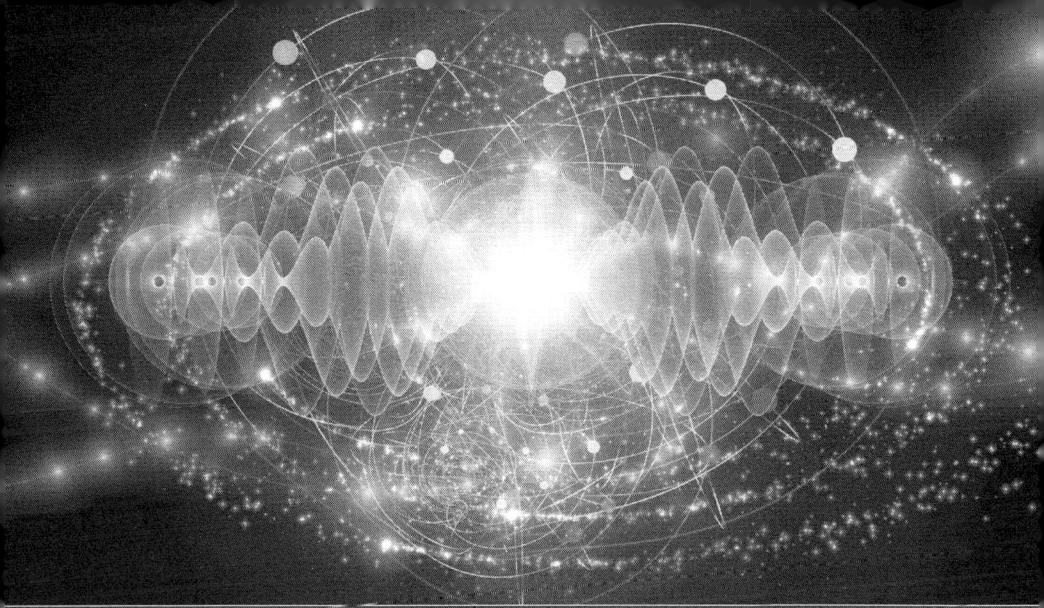

# TABLE OF CONTENTS

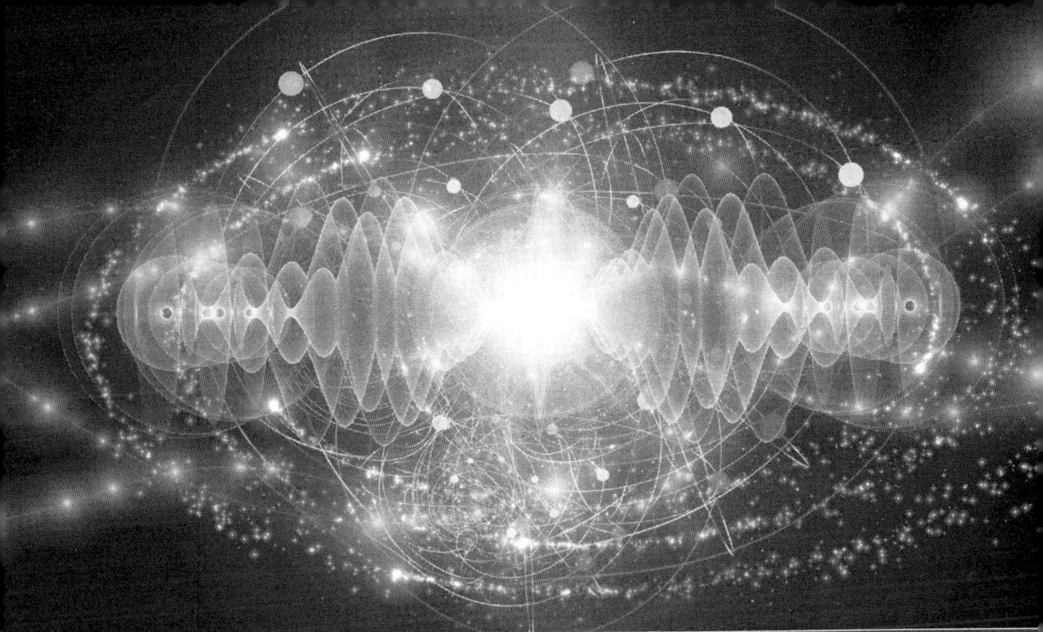

# PREFACE

Light has always fascinated mankind and since the beginning of recorded history it has been both a subject of research and a tool for investigation of other phenomena. Today, with the advent of nanotechnology, the use of light has reached its own dimension where light-matter interactions take place at wavelength and subwavelength scales and where the physical/chemical nature of nanostructures controls the interactions. This is the field of nanophotonics which allows for the exploration and manipulation of light in and around nanostructures, single molecules, and molecular complexes. Nanophotonics deals with the interaction of light with matter at a nanometer scale, providing challenges for fundamental research and opportunities for new technologies, encompassing the study of new optical interactions, materials, fabrication techniques, and architectures, including the exploration of natural and synthetic, or artificially engineered, structures such as photonic crystals, holey fibers, quantum dots, subwavelength structures, and plasmonics. Nanostructures have attracted considerable attention for many new fantastic applications owing to their remarkable physical properties. Most recent work has demonstrated that various material properties can be modified when the materials scale down to nanometers. Interacting with electromagnetic waves, nanomaterial's show exciting and interesting phenomena such as subwavelength wave guiding and modulation, light tripping and filtering, and ultra-resolution imaging. Although complete control of nanomaterial fabrication remains a challenge, various advanced techniques have been proposed to achieve sub-10 nm patterning. On this basis, desired radiation can be realized by manipulating the nanostructures. These nanostructures, especially the plasmonics nanostructures supporting surface plasmon polarities, show great potentials in many application fields. For these reasons, this research field, especially plasmonics and nanophotonics, has been experiencing an explosive growth.

This book will focus on all aspects of the fabrication and application of nanostructures and devices that help to understand or harness and utilize the behavior of light. Nanophotonics focuses on the interaction of photons with nano-structures, such as carbon nano-tubes, nano metal particles, nano crystals, semiconductor nano dots, photonic crystals, tissue and DNA. The book covers the latest developments for physicists, engineers and material scientists, working in fields related to: Plasmonics: metallic nanostructures and their optical properties Meta materials, fundamentals and applications. The book cover a wide variety of scopes in those areas including principles of plasmonics, SPR, LSPR and their applications, graphene-based nanophotonic devices, generation of entangled photon and quantum dots, perovskite solar cells, photo-detachment and photoionization of two-electrons systems, diffusion and intermixing of atoms in semiconductor crystals, lattice and molecular elastic and inelastic scattering including surface-enhanced Raman Scattering and their applications. The book deals with areas of physics, chemistry, and engineering of nanomaterials and their interactions with light leading to devices with photonics applications. It will consider manuscripts dealing with topics ranging from the design, creation, and experimental measurements of nanomaterials with novel and potentially useful optical properties to the testing and applications of nanophotonic devices.

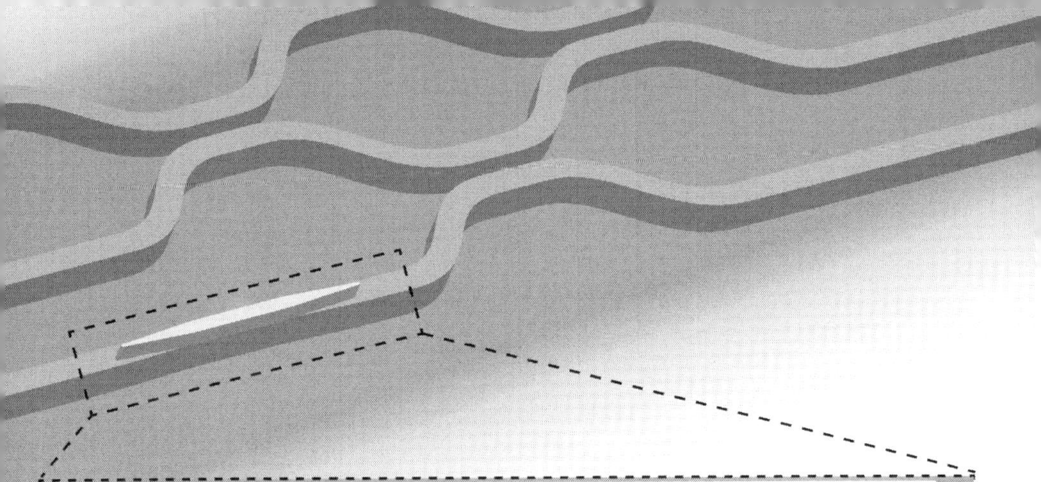

CHAPTER 1

# FUNDAMENTALS OF PLASMONICS

## INTRODUCTION

Plasmonics deals with the interaction of electromagnetic fields (i.e. light) with metallic structures. Typically, these metallic structures possess sub-wavelength dimensions and support the so-called surface plasmon resonances (SPR). These resonances occur due to the collective coherent oscillations of the free electrons present in the metal. SPR resonances are supported by thin metal films (propagating surface plasmons) or by metallic nano-structures (localised surface plasmons).

Plasmonics concerns to the investigation of electron oscillations in metallic nanostructures and nanoparticles (NPs). Surface plasmons have optical properties, which are very interesting. For instance, surface plasmons have the unique capacity to confine light at the nanoscale . Moreover, surface plasmons are very sensitive to the surrounding medium and the properties of the materials on which they propagate. In addition to the above, the surface plasmon resonances can be controlled by adjusting the size, shape, periodicity, and materials nature. Indeed, the technological progress allows researchers to produce new plasmonic systems by controlling all the parameters described previously. Moreover, theoretical, computational, and numerical simulation tools have been developed in this last decade, allowing for a better understanding of the optical properties of plasmonic systems. Thus, all these optical properties of plasmonic systems can enable a great number of applications, such as biosensors, optical devices, and photovoltaic devices.

## 1.1 ELECTRO-MAGNETICS IN METALS AND ON METAL SURFACES

The interaction of metals with electro-magnetic fields can be completely described within the frame of classical Maxwell equations:

$$\nabla \cdot \mathbf{D} = \rho \quad (1)$$

$$\nabla \cdot \mathbf{B} = 0 \quad (2)$$

$$\nabla \times \mathbf{E} = -\partial \mathbf{B}/\partial t \quad (3)$$

$$\nabla \times \mathbf{H} = \mathbf{J} + \partial \mathbf{D}/\partial t, \quad (4)$$

which connects the macroscopic fields (dielectric displacement D, electric field E, magnetic field H and magnetic induction B) with an external charge density ϱ and current density J.

In the limit of linear, isotropic and non-magnetic media, there are additionally the material dependent relations:

$$\mathbf{D} = \epsilon_0 \epsilon \mathbf{E} \quad (5)$$

$$\mathbf{B} = \mu_0 \mathbf{H} \quad (6)$$

with a frequency dependent dielectric constant: $\epsilon = \epsilon(\omega)$, which is in general a complex function, $\epsilon = \epsilon' + i\epsilon''$. It is furthermore connected to the complex index of refraction via $n = n + i\kappa = \sqrt{\epsilon}$. Explicitly one can obtain the following expressions:

$$\epsilon' = n^2 - \kappa^2, \qquad \epsilon'' = 2n\kappa, \tag{7}$$

$$n^2 = \frac{\epsilon'}{2} + \frac{1}{2}\sqrt{\epsilon'^2 + \epsilon''^2}, \qquad \kappa = \frac{\epsilon''}{2n} \tag{8}$$

The real part of the refractive index n(ω) is responsible for the dispersion in the medium, the imaginary part κ(ω) (extinction coefficient) determines the absorption. Beer's law describes the exponential decay of the intensity of a light beam (along the x direction) in a medium: $I(x) = I_0 \exp(-\alpha x)$. The absorption constant can then be determined from the extinction coefficient: $\alpha(\omega) = 2\kappa(\omega)\omega/c$.

### 1.1.1 Bulk Plasmons

The optical properties can be described over a large frequency range using the plasma model, where an electron gas (effective electron mass M) of density N is assumed that freely propagates behind a background of positively charged atom cores. These electrons start to oscillate in the presence of an electromagnetic field $E(t) = E_0 \exp(-i\omega t)$ and are damped through collisions with a characteristic rate $\gamma = 1/\tau$ (typically $\tau \approx 10^{-14}$ s at room temperature). The equations of motion in this model are

$$m\ddot{\mathbf{x}} + m\gamma\dot{\mathbf{x}} = -e\mathbf{E}(\mathbf{t})$$

with the solution

$$\mathbf{x}(\mathbf{t}) = \frac{e}{m(\omega^2 + i\gamma\omega)}\mathbf{E}(\mathbf{t})$$

The electrons which are displaced relative to the atom cores then generate a polarisation P = −Nex. From this it follows for the dielectric displacement and the dielectric constant:

$$\mathbf{D} = \epsilon_0\mathbf{E} + \mathbf{P} = \epsilon_0\epsilon\mathbf{E} \tag{9}$$

$$= \epsilon_0\left(1 - \frac{\omega_p^2}{\omega^2 + i\gamma\omega}\right)\mathbf{E} \tag{12}$$

$$\Rightarrow \epsilon(\omega) = 1 - \frac{\omega_p^2}{\omega^2 + i\gamma\omega} = 1 - \frac{\omega_p^2\tau^2}{\omega^2\tau^2 + i\omega\tau}, \tag{10}$$

where we introduced the plasmon frequency $\omega_p^2 = Ne^2/(\epsilon_0 m)$. In case of low frequencies $\kappa\tau \ll 1$ metals are strongly absorbing. The absorption constant then becomes $\alpha = \sqrt{2\omega_p^2\omega\tau/c^2}$. The penetration depth of the fields at low frequencies after Beer's law becomes $\delta = 2/\alpha$ and is called skin depth.

However at large frequencies the approximation $\omega\tau \gg 1$ is valid. In this case the damping term $i\omega\tau$ can be neglected and $\epsilon(\omega)$ becomes approximately real:

$$\epsilon(\omega) = 1 - \frac{\omega_p^2}{\omega^2} \tag{11}$$

The dispersion relation of electro-magnetic fields can be determined from $k^2 = |\mathbf{k}|^2 = \epsilon\omega^2/c^2$:

$$\omega(k) = \sqrt{\omega_p^2 + k^2/c^2}$$

As can be also seen in the figure 1, there is no propagation of electro-magnetic waves below the plasmon frequency $\omega < \omega_p$. For $\omega > \omega_p$ waves propagate with a group velocity $v_g = d\omega/dk < c$.

Of further interest is the special case $\omega = \omega_p$, where for low damping $(\omega_p) = 0$. One can show, that here a collective longitudinal excitation mode ($k \,||\, E$) is formed, with a purely de-polarizing field $(\mathbf{E} = (-1/\epsilon_0)\mathbf{P})$.

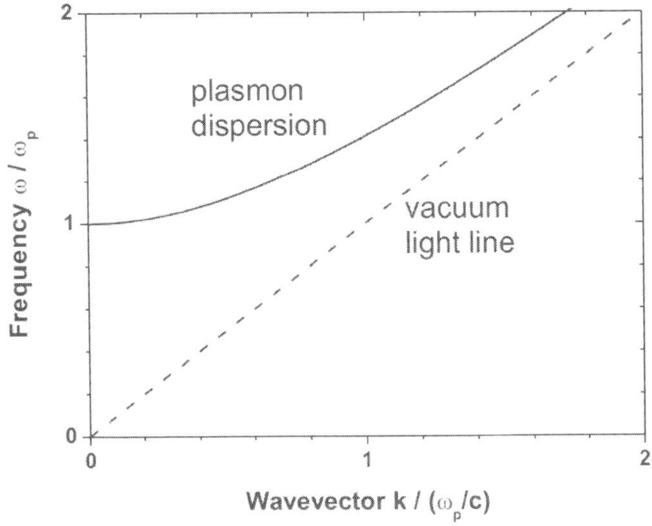


**Figure 1:** Dispersion relation of the free electron gas.

The physical interpretation is a collective oscillation of the conduction electron gas with respect to the fixed background of positive atom cores. The quanta of this charge oscillation are called plasmons (or bulk plasmons, for better discrimination with surface plasmons in the later sections). As these are longitudinal waves, bulk plasmons cannot couple to transversal electro-magnetic fields and thus cannot be excited from or strayed to direct irradiation.

In most metals, the plasma frequency is in the ultra-violet regime, with energies within 5-15 eV, depending on the metal band structure.

### 1.1.2 Surface Plasmons

Surface plasmons (or more exactly surface plasmon polaritons, SPPs) are electromagnetic excitations that propagate along the interface between a metal and a dielectric medium.

For the derivation of these excitation we again start with Maxwell's equations, which have to be separately solved for the metal and dielectric parts. Let us first start with a metal surface that extends infinitely in the x–y plane at z = 0 (see figure 2)

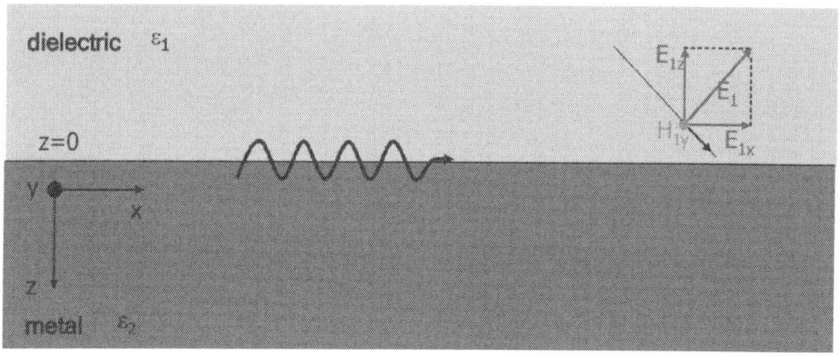


**Figure 2:** interface along the x–y plane between a dielectric (top, index 1), and a metal (bottom, index 2).

The conditions for the continuity of the normal and transversal field components on this interface are given by:

$$D_{1,z} = D_{2z}, \qquad B_{1,z} = B_{2,z} \tag{12}$$

$$E_{1,x/y} = E_{2x/y}, \qquad H_{1,x/y} = H_{2,x/y} \tag{13}$$

where the indices (1) and (2) indicate the dielectric and metal, respectively.

One can show, that under these conditions, no transverse-electric (TE) modes can exist. Instead we directly start with an ansatz for a transverse-magnetic (TM) mode for field which propagate along the x direction (i = 1, 2):

$$\mathbf{E}_i = (E_{i,x}, 0, E_{i,y})e^{i(\mathbf{k}_i \cdot \mathbf{r} - i\omega t)} \tag{14}$$

$$\mathbf{H}_i = (0, H_{i,y}, 0)e^{i(\mathbf{k} \cdot \mathbf{r} - i\omega t)} \tag{15}$$

$$\mathbf{D}_i = \epsilon_0 \epsilon_i \mathbf{E}_i, \qquad \mathbf{B}_i = \mu_0 \mathbf{H}_i \tag{16}$$

The wave vector is given by $k_i = (\beta, 0, k_{i,z})$, where $\beta = k_x$ indicates the propagation constant along x.

Using this approach with the conditions of continuity from above and the Maxwell's equation (31) and (34) in absence of charges and currents ($\varrho = 0$, $J = 0$), we obtain the following relations between the $k_{i,z}$ components:

$$\frac{k_{1,z}}{\epsilon_1} = \frac{k_{2,z}}{\epsilon_2} \qquad (16)$$

We are looking for solutions that describe modes bound to the interface. Thus, the $k_{i,z}$ components have to be imaginary and of opposite sign:

$$k_{1,z} = +i\kappa_1$$

In this way, the fields decay exponentially into the respective half spaces: $E_i \propto \exp(\pm ik_{i,z}) = \exp(\pm\kappa_{iz})$, as also symbolized in figure 3. Comparing this with the relation (16), one can directly see that this is fulfilled only, if the dielectric constants of the two materials are of opposite sign $(\text{i.e. } \epsilon_1 = -\epsilon_2)$. Surface plasmons can thus indeed only exist at the interface between a metal $(\epsilon < 0)$ and a dielectric medium $(\epsilon < 0)$.

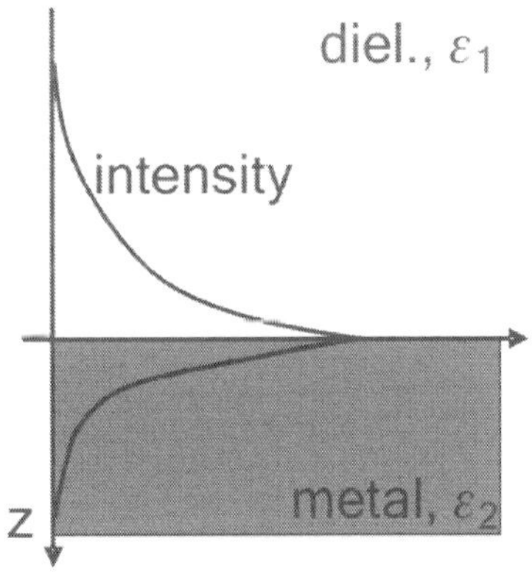


**Figure 3:** Evanescent field that decays exponentially into the two half spaces.

Altogether, one obtains a system that is composed of an electromagnetic wave in the dielectric medium and an oscillating electron plasma in the metal, where both modes have an exponentially decaying evanescent character (see figure 4). Due to this composed character, surface plasmons are also referred to as surface plasmon polaritons. The penetration depth of the field into the dielectric is typically on the order of $\lambda/2$ of the wavelength in the medium, whereas in the metal it is characteristically given by the skin depth.

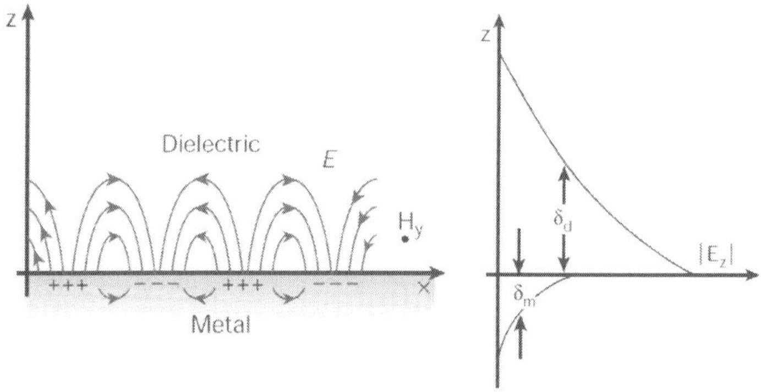


**Figure 4:** Left: composed character of SPPs at the interface between dielectric and metal. Right: evanescent fields in the two half spaces.

Let us now derive the dispersion relations for surface plasmons. For the wave vector from eq. (14) we get:

$$|\mathbf{k}_{1(2)}|^2 = \epsilon_{1(2)} k_0^2 = \beta^2 + k_{1(2),z}^2 = \beta^2 - \kappa_{1(2)}^2$$

here, $k_0 = \omega/c$ is the vacuum wave vector of light with frequency $\omega$. Insertion of $\kappa_{1(2)}$ into eq. (16) yields the propagation constant:

$$\beta = \frac{\omega}{c}\sqrt{\frac{\epsilon_1\epsilon_2}{\epsilon_1 + \epsilon_2}} \quad (17)$$

Together with the information of the frequency dependence of $\epsilon_{1(2)}(\omega)$, one can derive for the dispersion relation $\omega(k_{SPP})$.

Under the assumption that $\epsilon_1(\omega)$ in the dielectric is approximately constant and using the approximation (11) for $\epsilon_2$ in the metal above the plasmon frequency, one obtains a dispersion relation as given in figure 5.

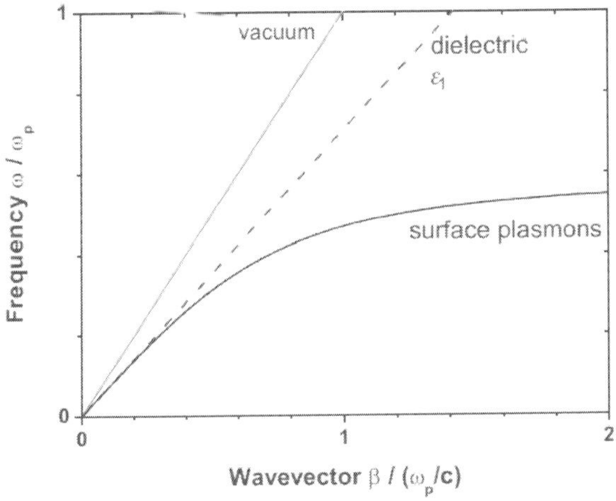


**Figure 5:** Dispersion relation of surface plasmons compared to light in vacuum and in the dielectric medium.

One realizes that the SPP dispersion relation completely resides below the light cones, while a crossing only occurs in the limit $\omega = 0$, $\beta = 0$. SPPs thus cannot be excited by direct illumination of light, as energy and momentum conservation ($\omega_{light} = \omega_{SPP}$ and $k_{light} = k_{SPP}$) can not be fulfilled at the same time. Instead, in order to excite SPPs, a momentum transfer has to be established.

In the above discussion we have not yet considered the complex nature of the dielectric constant of metal. Figure 6 shows the real and imaginary part of the dielectric constant for some selected metals. The imaginary part of the dielectric constant causes Ohmic damping of the electrons oscillations in the metal. When using eq. (17), one sees that the propagation constant also becomes imaginary, $\beta = \beta' + i\beta''$. As a consequence, in the solutions for the SPPs, eq. (14), an exponentially decaying damping term $\exp(-\beta''x)$, which leads to a characteristic propagation length of SPPs is given by

$$\delta_{SP} = \frac{1}{2\mathrm{Im}(\beta)}$$

These typically range within a few micrometers at visible wavelength and up to 100μm in the infrared.

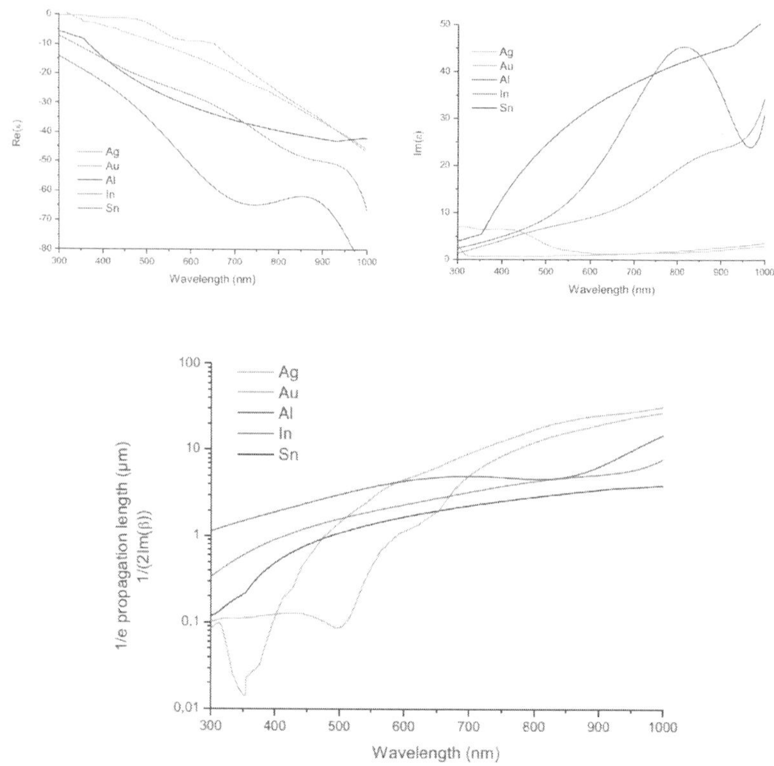


**Figure 6:** Real (left) and imaginary (right) part of the dielectric constant for some selected metals. Bottom: propagation lengths.

This situation becomes different in case of metal films (thickness d) with two interfaces to dielectric substrates. In the case of a thick metal film on a dielectric substrate two independent SPP modes exist related to different dielectric constants of the media adjacent to metal interfaces. These modes will degenerate if the film is in a symmetrical environment. If a metal film is thin enough so that the electromagnetic interaction between the interfaces cannot be neglected, the SPP dispersion is significantly modified and coupling between SPP modes on different interfaces of the film must be considered. The interaction of SPP modes removes the degeneracy of the spectrum and the surface plasmon frequencies split into two branches corresponding to symmetric (low-frequency mode) and antisymmetric (high-frequency mode) field distributions through the film. For large SPP wavevectors these modes can be estimated from

$$\omega_{\pm} = \omega_{SP}\sqrt{1 \pm \exp(-\beta d)}$$

At the same time, for such thin metal films, the imaginary part of the wavevector determining the propagation length of high-frequency SPP modes decreases as $d^2$ with the film thickness. This leads to a very long propagation length of such SPP modes, called the long-range SPPs. These modes can have propagation lengths up to centimeters in the infrared regime.

## 1.2 EXCITATION OF SURFACE PLASMON POLARITONS

The SPP dispersion curve lies entirely below that of free space light in the dielectric, such that $\beta > k$. Direct excitation of SPPs by light beams is thus not possible unless special techniques to achieve phase matching are employed.

### 1.2.1 Prism Coupling

This coupling scheme, also known as attenuated total internal reflection, involves the coupling of the SPPs to the evanescent electro-magnetic field that is formed upon total internal reflection of a light beam at a surface in an optically dense medium. Two different geometries for prism coupling are possible, see figure 7.

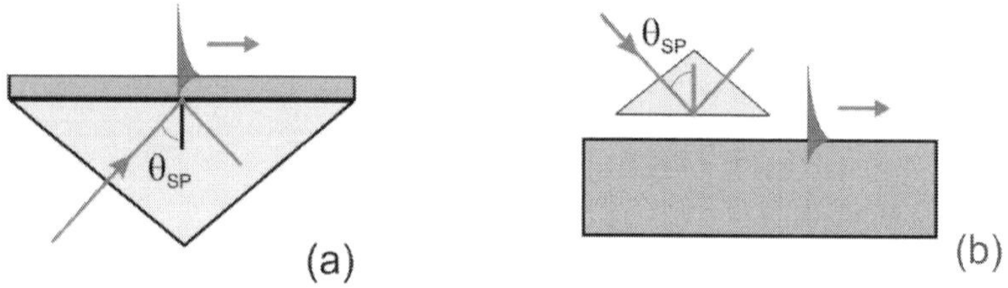


**Figure 7:** Prism coupling: (a) Kretschmann, (b) Otto configuration.

In the Kretschmann configuration, the metal film is evaporated on top of a glass prism. The film is illuminated through the dielectric prism at an angle of incidence greater than the angle of total internal reflection. The wavevector of light is increased in the optically dense medium. At a certain angle θ of incidence where the in-plane component of the photon wavevector in the prism coincides with the SPP wavevector on an air–metal surface, resonant light tunnelling through the metal film occurs and light is coupled to the surface polaritons:

$$\beta = \frac{\omega}{c}\sqrt{\epsilon_{\text{prism}}}\sin\theta$$

Under these resonant conditions, a sharp minimum is observed in the reflectivity from the prism interface as light can be coupled to SPPs with almost 100% efficiency. With the increase in the metal film thickness the efficiency of the SPP excitation decreases as the tunnelling distance increases. SPP on an interface between the prism and metal cannot be excited in this geometry as the wavevector of SPP at this interface is greater than the photon wavevector in the prism at all angles of incidence. To be able to excite SPP on the internal metal interface, an additional dielectric layer with a refractive index smaller that of the prism should be deposited between the prism and the metal film. In such a two-layer geometry, the photon tunnelling through this additional dielectric layer can provide resonant excitation of SPP on the inner interface. Thus, both SPP modes (on the surface and the interface) can be excited in such a configuration at different angles of illumination.

For thick metal films (or surfaces of bulk metal), for which the Kretschmann configuration cannot be used, SPP can be excited in the Otto configuration. Here, the prism where total internal reflection occurs is placed close to the metal surface, so that photon tunnelling occurs through the air gap between the prism and the surface. The resonant conditions are analogous to those in the Kretschmann configuration. This configuration is also preferable when direct contact with the metal surface is undesirable, e.g. for studying the surface quality.

### 1.2.2 Grating Coupling

The mismatch in wavevector between the in-plane momentum $k_x = k \sin \theta$ of impinging photons and β can also be overcome by using diffraction effects at a grating pattern on the metal surface. For a one-dimensional grating of grooves with lattice constant a, as depicted in figure 180, phase-matching takes place whenever the condition

$$\beta = k \sin \theta \pm nG$$

is fulfilled, where $G = 2\pi/a$ is the reciprocal vector of the grating and n = 1, 2, 3, . . .. As with prism coupling, excitation of SPPs is detected as a minimum in the reflected light.

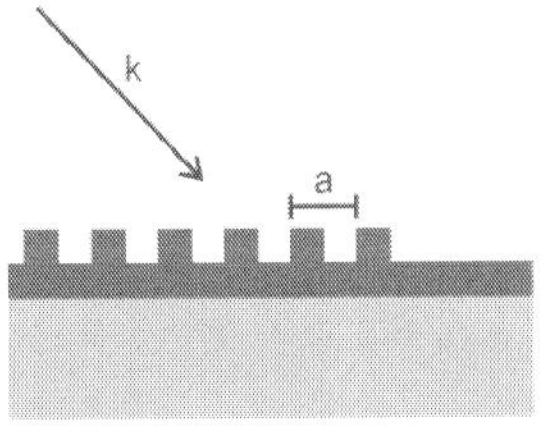


**Figure 8:** Grating coupling of a light with wave vector k impinging on a metal grating surface of period a.

The reverse process can also take place, when SPPs propagating along a modulated surface can couple to light and thus radiate (see figure 9).

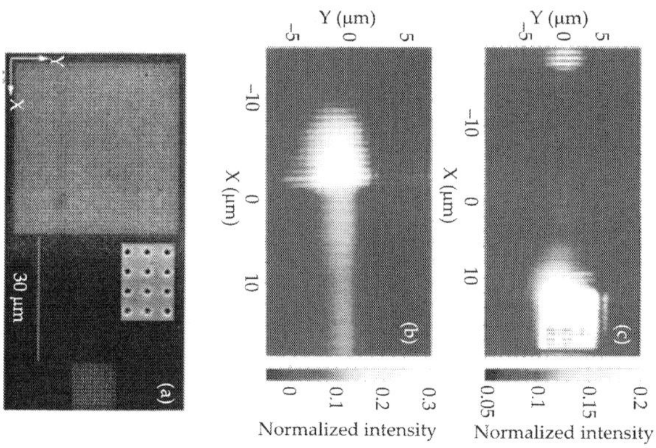

**Figure 9:** Left: SEM image of the pattern. 2 arrays with a period of 760 nm are separated by 30 mm. The hole diameter is 250 nm. The small array at the bottom is the source array whereas the big array at the top is not illuminated by the laser light and acts as a probe array. Inset: zoom in on the holes of one array. Right: Near-field optical image of the pattern when the laser is focused on the small array and the electric field is along the x direction.

For one-dimensional gratings, significant changes to the SPP dispersion relations occur if the gratings are sufficiently deep so that the modulation can not longer be treated as a small perturbation. An SPP photonic bandgap is then formed when the period is equal to half the wavelength of the SPP. Just as for electron waves in crystalline solids, there are two SPP standing wave solutions, each with the same wavelength but, owing to their different field and surface charge distributions, they are of different frequencies (see figure 10(a)). The upper frequency solution ω+ is of higher energy because of the greater distance between the surface charges and the greater distortion of the field. SPP modes with frequencies between the two band edges cannot propagate, and so this frequency interval is known as a stop gap.

### 1.2.3 Excitation using Highly Focused Optical Beams

As a variant of the prism coupling technique, a microscope objective of high numerical aperture can be used for SPP excitation.

An oil immersion objective is brought into contact with the glass substrate of a thin metal film via a layer of index matched immersion oil.

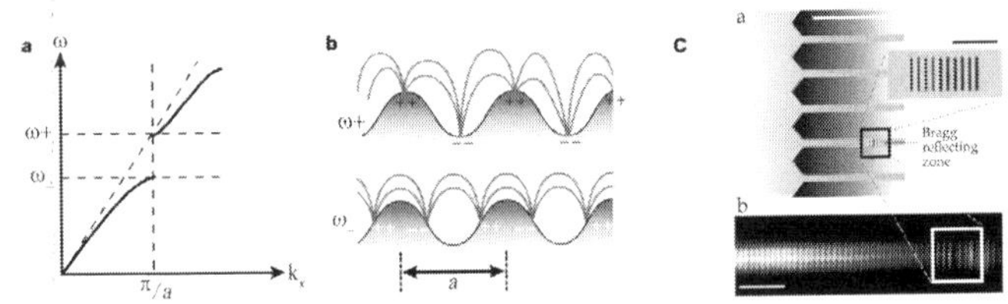

**Figure 10:** (a) SPP dispersion relation of a periodically modulated surface. (b) Schematic represantation of the SPP modes near the band edges. (c) Realization of a SPP Bragg reflector.

Brought into contact with the glass substrate of a thin metal film via a layer of index matched immersion oil. The high N.A. of the objective ensures a large angular spread of the focused excitation beam, including angles $\theta > \theta_c$ greater than the critical angle for total internal reflection. In this way, wave vectors $k_x = \beta$ are available for phase-matching to SPPs at the metal/air interface. Off-axis entrance of the excitation beam can further ensure an intensity distribution preferentially around $\theta_{SPP}$ thus reducing the amount of directly transmitted/reflected light. The highly focused beam also allows for localized excitation within the diffraction limit. This scheme is especially convenient for the excitation of a continuum of SPPs at different frequencies. and the determination of their propagation lengths.

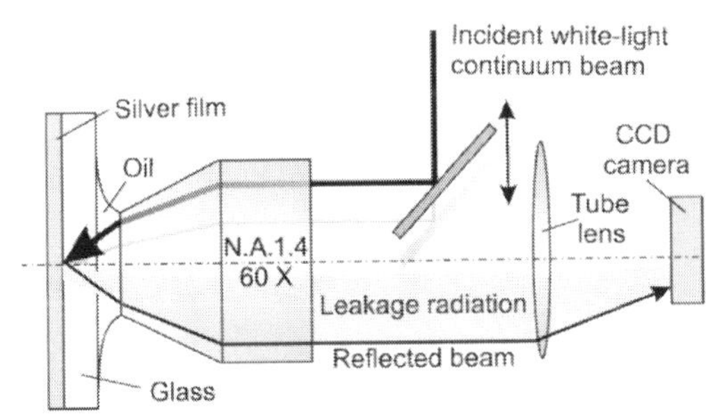


**Figure 11:** Schematic of the excitation of a white-light continuum of SPPs and their observation through an index-matched oil-immersion lens.

### 1.2.4 Near-field Excitation

Near-field applications in SPP studies brought forth a new technique which provides the possibility to excite SPP locally with sub-wavelength resolution. Additionally, the effect of surface roughness on the SPP propagation and the scattering at individual surface defects can be studied with high spatial resolution.

Using illumination through a SNOM fiber tip, circular SPP waves can be locally launched at the surface. This configuration (see figure 12) can be treated as either a diffraction or tunnelling mechanism of the SPP excitation. In the former case, one can consider a near-field coupling of the light diffracted on the sub-wavelength aperture of the fiber tip in SPPs. In the latter case, because mainly the evanescent field components are generated at the aperture, one can describe the process as the photon tunnelling taking place from the fiber tip to the metal surface. This technique is somewhat analogous to the Otto configuration but allows local (with sub-wavelength precision) excitation of SPPs at the SNOM tip position.

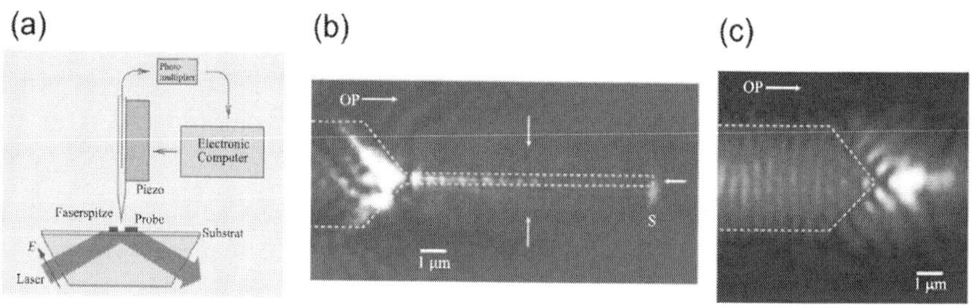


**Figure 12:** (a) Principle of the detection of SPPs (excited via the prism) with a SNOM probe. (b) and (c) SNOM images, propagation of SPPs along gold nanowires (wavelength 633 nm).

### 1.2.5 Coupling to Integrated Photonic Elements

While the excitation schemes above are interesting for investigation of SPP propagation, practical applications of SPPs will require compact and efficient coupling schemes in integrated photonic circuits. Preferably, the plasmonic components should allow efficient matching with conventional dielectric waveguides and fibers.

One such coupling scheme is end-fire coupling, in which a free-space optical beam is focused on the end-facet of the desired waveguide. This scheme operates via matching the spatial field distribution of the waveguide as much as possible by adjusting the beam width. This can be further miniaturized by directly connecting an optical fiber to the end-faced of the plasmon waveguide (see figure 13).

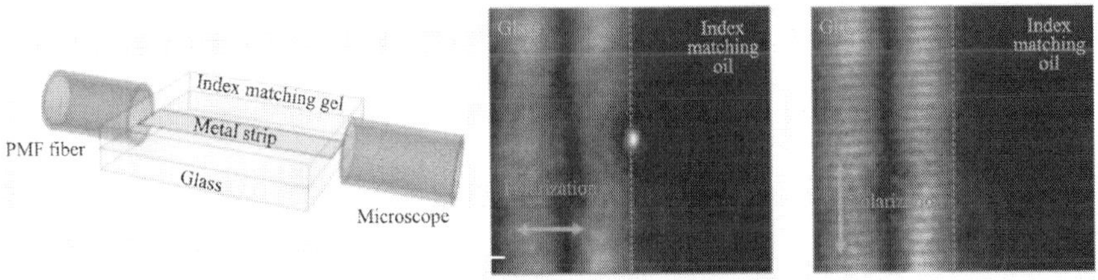


**Figure 13:** A polarization maintaining fiber coupled to a HeNe laser illuminates one end of the strip and the microscope collects the light scattered at the other end. The strip output lights up when the electric field is parallel to the short axis of the strip. This result is as expected because the SPP modes are mostly TM polarized.

A convenient interfacing scheme makes use of optical fiber tapers brought into the immediate vicinity of the waveguide to achieve phase-matched power transfer via evanescent coupling (see figure 14). Fiber tapers are standard glass optical fibers that have been heated and stretched to a diameter that is at or below the wavelength of light, at which point the evanescent field of the guided mode extends into the surrounding air and allows the taper to function as a near-field optic. Similar approaches are used to couple light to microcavity structures such as microdisks.

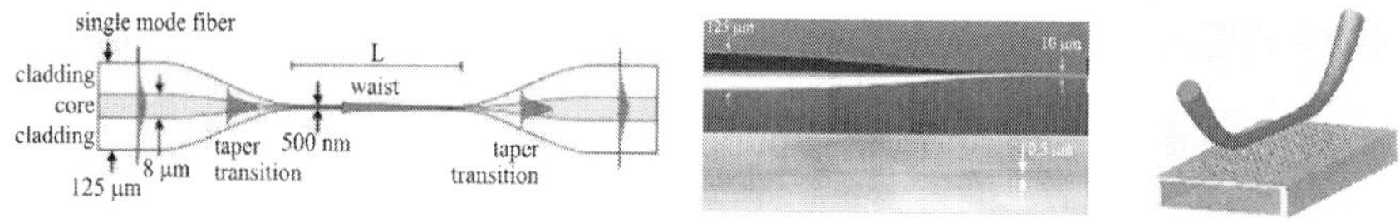


**Figure 14:** Left and center: Fiber tapers (D. Meschede, Univ. Bonn). Right: Fiber taper coupled to an SPP nanoparticle waveguide.

## 1.3 SURFACE PLASMON WAVEGUIDES

In this subsection we discuss how to control the propagation of SPPs along waveguiding structures. The main figures of merit are here the confinement of SPPs to small dimensions and small losses.

### 1.3.1 Waveguides for Long-Range SPP Modes

A very thin metal films of thickness t, that are sandwiched between two symmetric layers of dielectric, interactions between SPPs on the bottom and top interfaces lead to the occurrence of coupled modes. The odd modes displays the interesting property of dramatically decreased attenuation with reduction in the metal thickness. Waveguides can be designed by defining metal stripes of lateral width w along which the SPPs propagate.

First we will consider the case $w, \lambda \gg t$. In figure 15, the dispersion relation for the propagation constant β and the attenuation $\alpha$ of such a metal stripe is given for the diverse existing modes. The symmetric modes are denoted by $ss_b^0$ . One sees that the propagation constant dramatically decreases with reducing layer thickness. The long-range SPP modes can have show 1/e attenuation lengths up to the order of cm. This can by explained by the fact that only a small amount of the field resides in the thin metal layer, where Ohmic losses can occur. Instead the mode confinement increases to the order of the free-space wavelength into the dielectric. One thus has to find a compromise between small mode confinement and sub-wavelength guiding and, on the other hand, the minimization of losses.

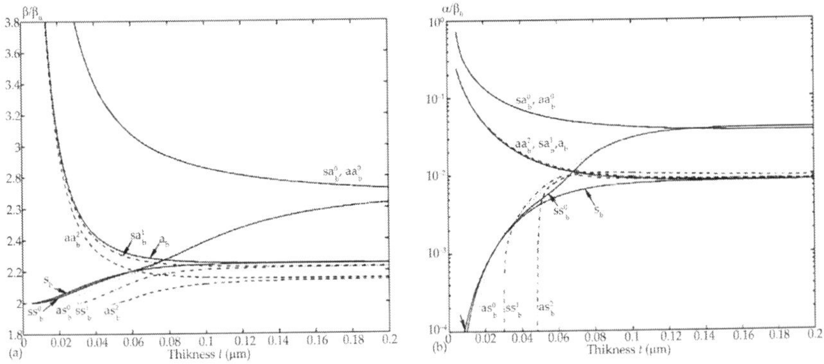


**Figure 15:** Dispersion relation for the propagation constant β and the attenuation α in a thin silver metal stripe of thickness t, width w = 1μm and infinite length, embedded in a dielectric with $\epsilon = 4$. The wavelength is 633 nm. The symmetric long-range SPP modes are denoted by $ss_b^0$ .

Figure 16 shows an example, of how such long-range SPP waveguides can be implemented to even design complex plasmon-optical elements, such as beamsplitters, interferometers, or Bragg gratings.

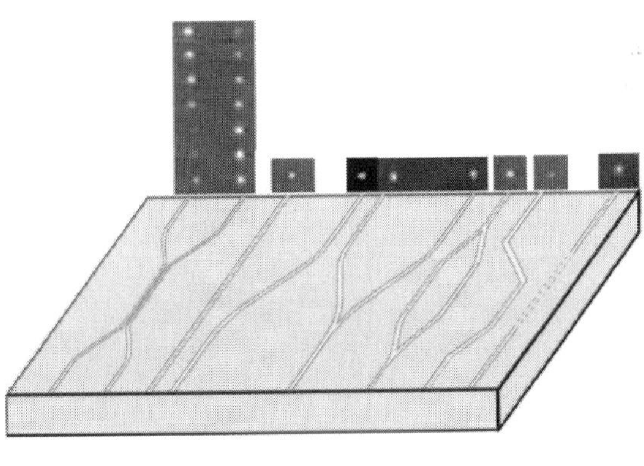

**Figure 16:** Waveguide structures for long-range SPPs. The gold stripes are 25 nm thick and have widths between 2–8 μm and are sandwiched between $SiO_2$ and a index-matching gel (both $\epsilon = 2.1$). The CCD images show the out-coupled light (λ = 1550 nm).

## 1.3.2 SPP Guiding in Metal Nanowires

In contrast to the previous chapter, we now discuss the case of metal wires of small sizes in both dimensions, $w, t \ll \lambda$.

It is instructive to compare this with dielectric waveguides. In a dielectric waveguide with $\epsilon_{core} > 0$ and for propagation along z, the relation between propagation constant β, transverse wave vector components $k_{x,y}$ and frequency is

$$\beta^2 + k_x^2 + k_y^2 = \epsilon_{core}\frac{\omega^2}{c^2}$$

Since $\epsilon_{core} > 0$ and $k_{x,y}$ are real, this implies that $\beta, k_{x,y} \leq \sqrt{\epsilon_{core}}\omega/c = 2\pi n_{core}/\lambda_0$. According to the uncertainty relation between wave vector and spatial coordinates, the mode size is thus limited by the effective wavelength in the core medium,

$$d_x, d_y \geq \lambda_0/(2n_{core}) \qquad (18)$$

However, if the guiding medium is a metal, then $\epsilon_{core} > 0$ and the transverse wave vectors kx,y must be imaginary. The guided waves are now one-dimensional. In this case, eq. (18) does not apply and the mode size can be substantially below the diffraction limit of the surrounding dielectric cladding.

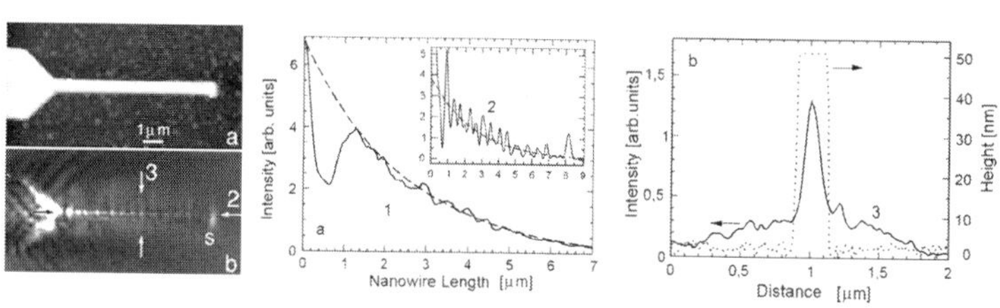


**Figure 17:** Left: topography (a) and near field intensity (b) of a gold nanowire (t = 50 nm, w = 200, λ0 = 800 nm, length: 8 µm). The nanowire is addressed by a 3 m wide gold stripe, the SPP is propagating from left to right. The arrows indicate the positions of the cross-cuts in the graphs. Center: cross-cut of the optical near-field intensity along the axis of the

nanowire. Dashed line: exponential fit with L = 2.5μm. Right: cross-cut of the optical near-field intensity perpendicular to the nanowire. The corresponding topography profile taken from a scanning electron microscope image is displayed by the dotted line.

The improved mode confinement goes along with increased Ohmic damping. An example for such a nanowire is given in figure 17. The longitudinal cross section of the near-field scan shows that indeed the propagation length is clearly smaller (several μm), than in the case of long-range SPPs, while the lateral cross section confirms the very strong confinement of the mode to the nanowire.

The propagation length in such systems can be drastically improved, however, when using chemically synthesized nanowires (CNW), instead of lithographically produced structures (LNW). In the latter case, the metal nanowires are produced by sputtering a gold film on a substrate and removing the spare parts using lithographic techniques. In this sputtering process produces a metal layer of poly-crystalline structure and shows some surface roughness that introduces further damping of the SPP by scattering. In contrast, CNWs can be fabricated with a well defined crystal structure, thereby minimizing surface plasmon damping due to scattering at roughness, domain boundaries, or defects.

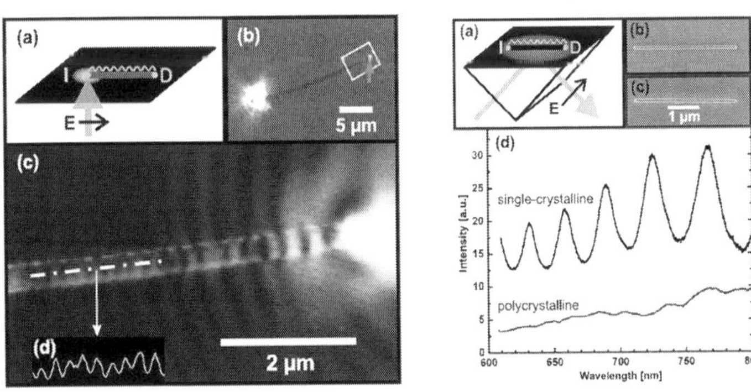


**Figure 18:** Left: (a) SPP propagation along the nanowire. I is input and D is distal end of the wire. (b) Microscopic image. The bright spot to the left is the focussed exciting light. The arrow indicates light scattered from the distal wire end. (c) SNOM image. (d) Cross-cut along the dotted line. Right: Scattered light spectra of 3.3 μm silver nanowires, diameter 90 nm. (a) Sketch of optical excitation. The exciting light propagation

direction projected onto the substrate plane is parallel to the nanowire axis, and the polarization is fixed in the plane of incidence. (b),(c) SEM images of a CNW and a LNW, respectively. (d) Scattered light spectra from the distal nanowire end facet of the CNW (single-crystalline) and the LNW (polycrystalline).

Figure 18 compares a CNW and a LNW (silver). The excited SPPs are expected to be reflected at the nanowire end facets and to form a standing wave. This can be indeed observed in the monocrystalline CNW, while the increased damping in the LNW hinders the formation of such standing waves.

### 1.3.3 Localized Modes in Gaps, Grooves, and Ridges

Coupled SPP modes in dielectric/metal/dielectric structures. Similar modes can also occur in inverse geometries, i.e. metal/dielectric/metal structures. Here we discuss some possible realizations. As shown in figure 19, different modes with orthogonal polarization are formed, that are either located inside the gap or at the edges of the metal layer, respectively. These modes have propagation lengths, in terms of power attenuation by 1/e, in the range of 1 mm in the infrared, offering a low losses SPP guiding, similar to Long-Range-SPP modes, supported by thin metal stripes.

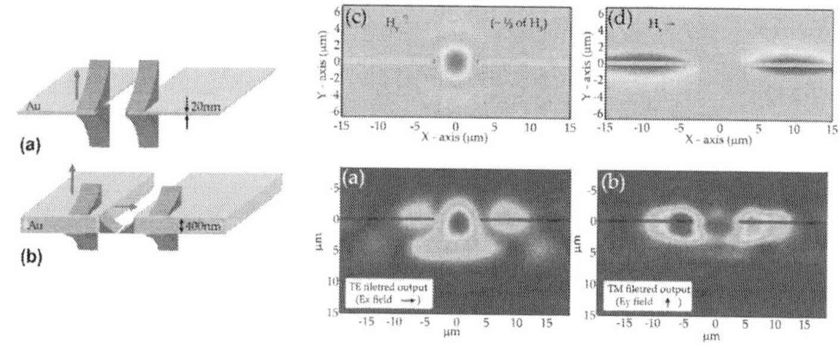


**Figure 19:** Modes in channel SPP waveguides. The structures are 400 nm gold layers on Si. The SPPs were excited at 1550 nm by end-fire coupling. The modal profile at the output facet were simulated (top images) and imaged through a high resolution objective onto a CCD camera (bottom images).

Similar modes can be defined by metal grooves and wedges. Examples of SPP propagation at $\lambda_0 = 1550$ nm are shown in figure 20 for different integrated plasmonoptical elements (beamsplitters, Mach-Zehnder interferometer, and ring interferometer). Depending on the groove width, the propagation lengths can be up to 100 µm or even longer, with the CPP field extending out of the groove.

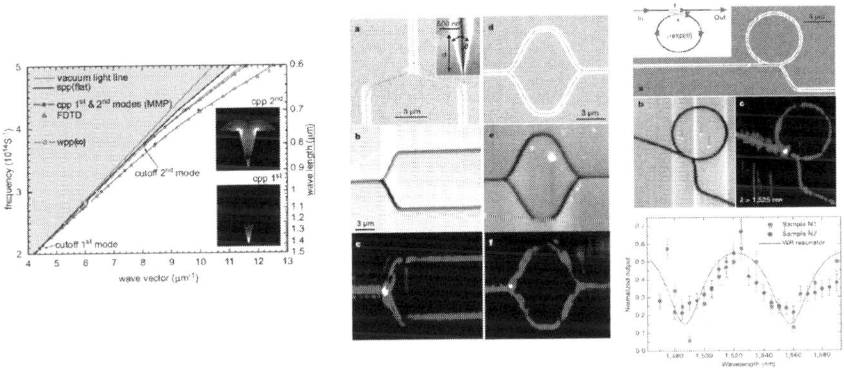


**Figure 20:** Left: dispersion relation and field distribution of different modes in groove SPP waveguides. Center: Plasmonic Y-splitter and MachZehnder interferometer (from top to bottom: SEM, topographic, and near field images). Right: SPP ring resonator. (a) SEM, (b) topographical, (c) near-field image. Bottom graph: transmission spectra for two ring resonators

An additional interesting feature of gap/groove waveguides is their lower susceptibility for losses at bent structures. While for the long-range SPP waveguides curvature radii of ~ 100µm or more are necessary, the radii used in figure 192 are on the order of 2 µm (≈ 1.5λ).

## 1.4 LOCALIZED PLASMONS ON METAL NANOPARTICLES

Here we introduce another fundamental plasmonic excitation, localized surface plasmons (LSPs) (also known as surface plasmons on metal nanoparticles). LSPs are non-propagating plasmon excitations on metallic nanostructures.

For small isolated metal particles with sizes in the range of the penetration depth of an electromagnetic field (e.g. ◎ 20 nm for Ag in the optical range), the clear distinction between surface and bulk plasmon vanishes. An external field can penetrate into the volume and shift the conduction electrons with respect to the ion lattice (see figure 21).

The coherently shifted electrons together with the restoring field represent an oscillator, whose behaviour is defined by the effective electron mass, charge density and geometry of the particle. Most physical effects associated with LSPs can be explained with this simple model. Spectrally the LSP resonances are in the visible to near infrared spectral regime. If excited at resonance, the amplitude of the induced electromagnetic field can exceed the exciting fields by a factor on the order of 10.

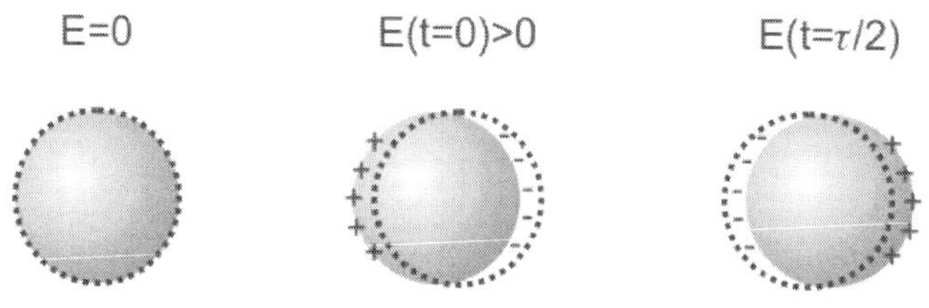


**Figure 20:** Localized surface plasmons on a metal nanoparticle in the absence (left) and presence (center & right) of an external electromagnetic field.

### 1.4.1 Plasmon Resonance of a Small Spherical Particle

Here we discuss in some more detail a LSP model in case of a small spherical particle. We assume a sphere of radius a in a uniform static electric field $E_{in} = E_0 e_z$ The surrounding medium is isotropic and non-absorbing with dielectric constant $^{2}$d, the metal sphere is described by a complex dielectric constat $\epsilon_s$. We are interested in a solution of the Laplace equation $\Delta\Phi = 0$ from which we can obtain the electric field E = −gradΦ. In spherical coordinates, the Laplace equation is of the form

$$\frac{1}{r^2 \sin\theta}\left[\sin\theta\,\partial_r(r^2\partial_r) + \partial_\theta(\sin\theta\,\partial_\theta) + \frac{1}{\sin\theta}\partial_\phi^2\right]\Phi(r,\theta,\phi) = 0$$

Due to the azimuthal symmetry, the general solution is independent of φ and has the form

$$\Phi_s(r,\theta) = \sum_{l=0}^{\infty} A_l r^l P_l(\cos\theta) \tag{19}$$

$$\Phi_d(r,\theta) = \sum_{l=0}^{\infty} (B_l r^l + C_l r^{-l-1}) P_l(\cos\theta) \tag{20}$$

where $\Phi_s$ is the potential inside the sphere and $\Phi_d = \Phi_{scatter} + \Phi_0$ the potential outside the sphere, consisting of an incoming and a scattered part.

When applying boundary conditions at the interface r = a for tangential part of the electric field ($\partial_\theta \Phi_i$ = const., for i = s, d) and for the longitudinal part of the displacement field $(\epsilon_i \partial_r \Phi_i = \text{const.})$, the potentials evaluate to

$$\Phi_s(r,\theta) = -\frac{3\epsilon_d}{\epsilon_s + 2\epsilon_d} E_0 r \cos\theta \tag{21}$$

$$\Phi_d(r,\theta) = -E_0 r\cos\theta + \frac{\epsilon_s - \epsilon_d}{\epsilon_s + 2\epsilon_d} E_0 \frac{a^3}{r^2}\cos\theta \tag{22}$$

$$= -E_0 r\cos\theta + \frac{\mathbf{p}\cdot\mathbf{r}}{4\pi\epsilon_0\epsilon_d r^3} \tag{23}$$

$$\text{with } \mathbf{p} = 4\pi\epsilon_0\epsilon_d a^3 \frac{\epsilon_s - \epsilon_d}{\epsilon_s + 2\epsilon_d}\mathbf{E}_0 \tag{24}$$

Here we have introduced the dipole moment p that is induced in the sphere by the external field. The electric field can then be expressed as

$$\mathbf{E}_s = \frac{3\epsilon_d}{\epsilon_s + 2\epsilon_d}\mathbf{E}_0 \tag{25}$$

$$\mathbf{E}_d = \mathbf{E}_0 + \frac{1}{4\pi\epsilon_0\epsilon_d}\frac{3\mathbf{n}(\mathbf{n}\cdot\mathbf{p}) - \mathbf{p}}{r^3} \tag{26}$$

n = r/r is the unit vector in direction of the point of interest.

The polarizability, defined by the relation $\mathbf{p} = \epsilon_0\epsilon_d\alpha\mathbf{E}_0$ then becomes

$$\alpha = 4\pi\epsilon_0 a^3 \frac{\epsilon_s - \epsilon_d}{\epsilon_s + 2\epsilon_d}$$

Note that this is of the same form as the Clausius-Mosotti equation. The scattering cross section of the sphere is then obtained by dividing the total radiated power of the dipole $(P = \omega^4/(12\pi\epsilon_0\epsilon_d c^3)|\mathbf{p}|^2)$ by the intensity of the exciting wave (I = $(1/2)c\epsilon_0\epsilon_d E_0^2$):

$$\sigma_{\text{scatt}} = \frac{k^4}{6\pi\epsilon_0^2}|\alpha(\omega)|^2 \tag{27}$$

$$= \frac{8\pi}{3}k^4 a^6 \left|\frac{\epsilon_s - \epsilon_d}{\epsilon_s + 2\epsilon_d}\right|^2 \tag{28}$$

Figure 21 shows plots of the normalized scattering cross-section for silver and gold particles. A redshift of the resonance is observed by increasing the dielectric constant of the dielectric environment.

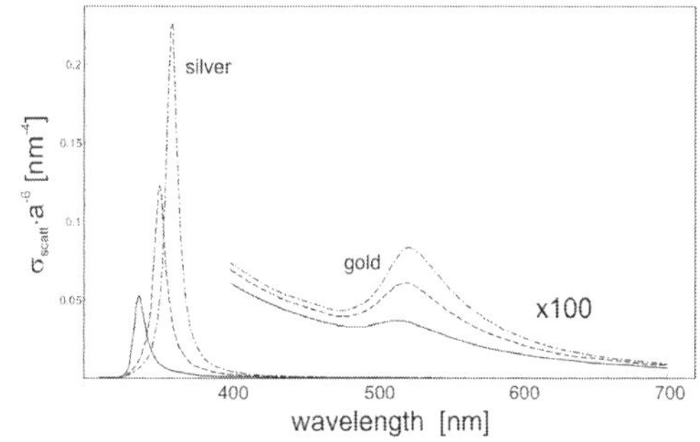


**Figure 21:** Plots of the scattering cross-section of spherical Ag and Au particles in different environments normalized by $a^6$ (a: particle radius).

Solid line: vacuum, n = 1, dashed line: water, n = 1.33, dash-dotted line: glass, n = 1.5.

The power removed from the incident beam due to the plasmon is not only due to scattering but also due to absorption. The sum of absorption and scattering is called extinction. Using Poynting's theorem, we know that the power dissipated by a point dipole is $P_{\text{abs}} = (\omega/2)\text{Im}(\boldsymbol{\mu} \cdot \mathbf{E}_0^*)$. Using $\boldsymbol{\mu} = \epsilon_d \alpha \mathbf{E}_0$, the absorption cross section becomes

$$\sigma_{\text{abs}} = \frac{k}{\epsilon_0}\text{Im}(\alpha(\omega)) \quad (29)$$

$$= 4\pi k a^3 \text{Im}\left(\frac{\epsilon_s - \epsilon_d}{\epsilon_s + 2\epsilon_d}\right) \quad (30)$$

One sees that the scattering scales with $a^6$ , while absorption process scale with $a^3$ . Consequently, for large particles scattering dominates, whereas for smaller particles, extinction is dominated by absorption. The transition between the two regimes is characterized by a color change. For example small gold particles absorb green and blue and thus appear as red. On the other hand, large gold particles mostly scatter in the green and appear greenish. This effect is used in color glasses.

## 1.5 APPLICATIONS OF SURFACE PLASMONS

There are some applications of surface plasmons:

### 1.5.1 Surface enhanced Raman scattering

An important application of surface plasmons is the enhancement of scattering of emitters near metal nanostructures. An example of this is Surface Enhance Raman Scattering (SERS)

The Raman effect describes the inelastic scattering between a photon and an emitter (e.g. a molecule) mediated by a vibrational or rotational mode of the emitter. Due to the energy exchange, the incoming photon of energy $\hbar\omega_L$ is shifted by the energy of the vibration $\hbar\omega_M$. If the photon looses energy, this is called Stokes scattering ($\omega_S = \omega_L - \omega_M$), if it gains energy by de-excitation of an vibrational mode it is called Anti-Stokes scattering ($\omega_{aS} = \omega_L + \omega_M$). This is symbolized in figure 22. In general the photons are not in resonance with the molecule but are excited to a virtual level.

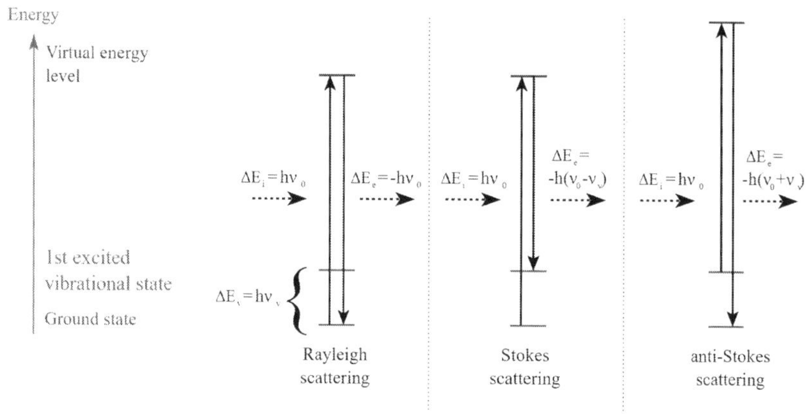


**Figure 22:** Schematics of the Raman scattering process.

Raman transitions can have very sharp lines and enable a detailed analysis of the molecule under study. Typically the Raman cross sections are more than 10 orders of magnitude smaller those of fluorescence processes ($\sigma_{Raman} \sim 10^{-31} - 10^{-29}$ cm$^2$/molecule).

SERS describes the enhancement of the Raman scattering by placing the molecule within the near-field of a metallic nanostructure (see figure 23) or a roughened metal surface. This enhancement is due to two effects: firstly, the Raman cross section is modified due to the changed environment. A much more important factor is the increased electromagnetic field due to excitations of LSPs and a crowding of the electric field lines (lightning rod effect) at the metal interface. The highest enhancements recorded to date are on the order of $10^{14}$, achieved on roughened silver surfaces.

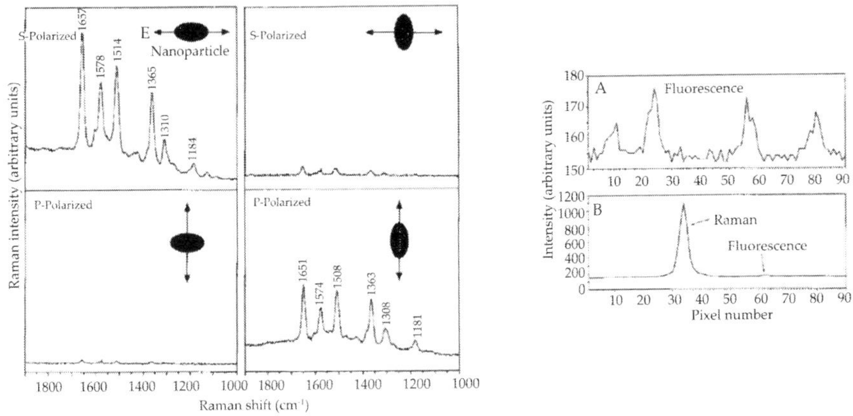


**Figure 23:** Left: SERS spectra of Rhodamine6G molecules on Ag nanoparticles (35 nm) with different orientation and polarization. Right: Comparison between the fluorescence and SERS signal.

### 1.5.2 Fluorescence Enhancement

Similar to SERS, the fluorescence of an emitter in the proximity of metal nanostructures can be enhanced as well. Fluorescence here results from excitation of the emitter by the incident field, which can show significant enhancement due to plasmon resonances in the metal particle. However, care has to be taken, in order not to quench the fluorescence due to non-radiative processes, such as absorption by the surface plasmon resonance. This allows the utilization of metal nanostructures as nano-antennas.

Figure 24 shows an experiment, in which a diamond nanocrystal is placed near a gold nanoparticle or sandwiched between two particles. The intensity of the fluorescence could be thus enhanced by a factor of ~ 6, and the radiative decay rate was increased by a factor 6–9. The single-photon character of the emitter was conserved in this process.

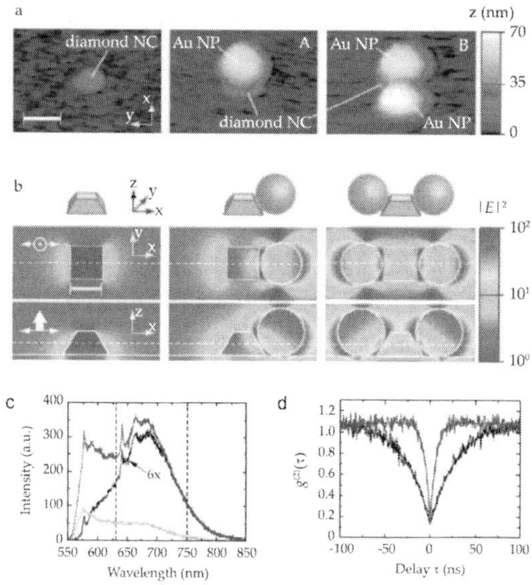


**Figure 24:** Fluorescence enhancement of a diamond nanocrystal (with nitrogen-vacancy defect center) near gold nanoparticles. (a) AFM images of three configurations: bare diamond, diamond near a Au particle, diamond between 2 Au particles. (b) Simulations of the intensity enhancement of the excitation light. (c) Fluorescence spectrum of the bare diamond (black, scaled by a factor of 6) and of configuration A under excitation parallel (blue) as well as perpendicular (orange) to the x axis at the same excitation intensity. (d) Normalized autocorrelation measurements on bare diamond (black) and configuration A (blue).

Figure 25 shows another experiment, in which terrylene molecules are excited and a gold nanoparticle attached to a scanning probe tip is moved over the sample with variable distances. This nicely demonstrates the sensitivity to the distances when coupling emitters to metal nanoparticles.

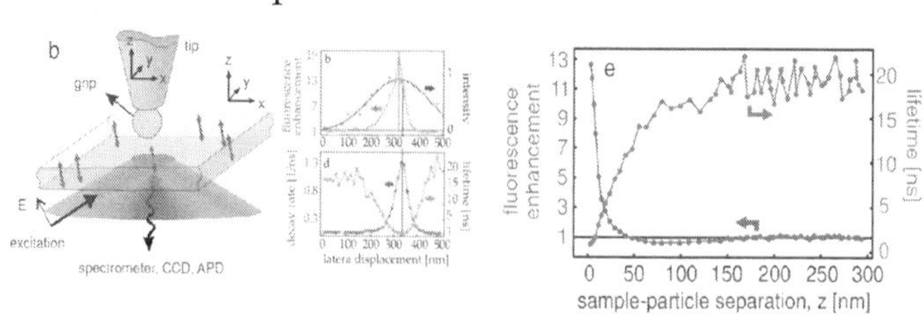


**Figure 25:** Left: Schematics of the experimental arrangement. Center: fluorescence enhancement (red), intensity (black), lifetime (brown), and

decay rate (blue) as a function of the lateral position of the scanned gold particle (100 nm) 1 nm above the sample. Right: Fluorescence enhancement and lifetime as a function of the vertical displacement of the Au particle.

Finally, figure 26 shows an experiment in which not only the decay rate of a nanocrystal quantum dot (QD) is enhanced, but which also shows that the photons from a single emitter can excite single surface plasmons in a metallic nanowire (NW). These single SPPs then penetrated along the NW and eventually scatter at its distal end into freely propagating single-photons. The correlation measurements at either end of the NW and the cross-correlation measurement between the direct particle emission and the scattered plasmon emission demonstrates that SPPs indeed originate from the emitter and that they act as bosons which obey the same photon number statistics as light.

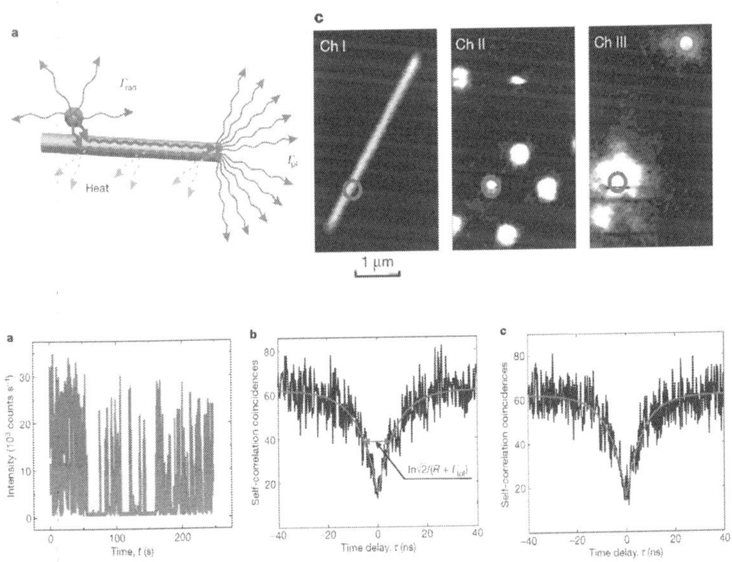


**Figure 26:** Top left: schematics of the experiment: a nanocrystal QD is placed next to a metal NW. Emission takes place in form of free photons as well as by exciting SPPs in the NW. Top right: μPL images of the system. I: nanowire image, II: image of QDs, III: the excitation laser was focused on the QD (red circle). The largest bright spot corresponds to the QD PL, while two smaller spots correspond to SPPs scattered from

the NW ends. The blue circle indicates the farthest end of the nanowire, used for photon cross-correlation measurements. Bottom: (a) Time trace of PL counts (red curve) from a coupled QD and scattered light (blue) from the end of the coupled NW. Fluctuations are due to QD blinking. (b) Second-order correlation function of QD PL. The number of coincidences at t = 0 goes almost to zero, confirming that the QD is a single-photon source. (c) correlation function between fluorescence of the QD and scattering from the NW end, obtained by coincidences between channel II (QD) and channel III (NW end).

### 1.5.3 Surface Plasmon Sensors in Biology and Medicine

Because the plasmonic interaction between metal nanoparticles is very sensitive to their separation and to the refractive index of the surrounding medium, they are used in biological, chemical, and medical sensing and detection applications.

The resonance shift due to a changing dielectric environment may allow for identifying different chemicals. In such an approach, a thin metal layer or surface containing a dense package of metal nanoparticles is prepared. When functionalizing the metal particles, they can be made sensitive to special types of bio-molecules. When the investigated chemical is rinsed over the surface, selective binding of species to the nanoparticles will change local dielectric properties resulting in a noticeable shift in LSP resonance (figure 200 right). Similarly, functionalized nanoparticles can be injected into cells and will bind to specific molecules, which can then efficiently be located (figure 200 left).

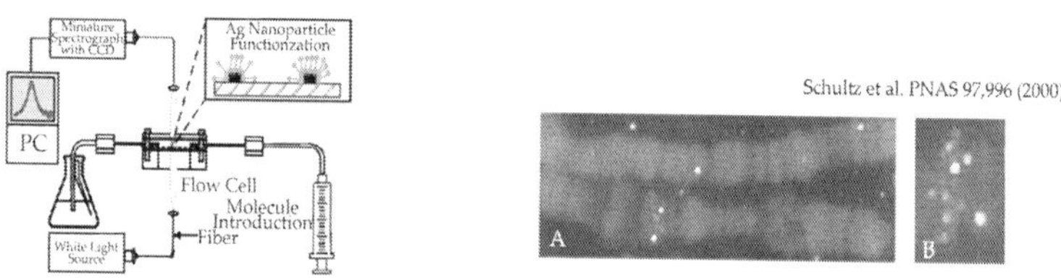


**Figure 27:** Left: sensing of bio-molecules by detecting the shift in LSP resonance. Right: specific binding of DNA coated nanoparticles to DNA can be used to locate certain complementary DNA sequences on chromosomes.

Because the plasmonic interaction between metal nanoparticles is very sensitive to their separation, precise measurements of the plasmon resonance wavelength of metal particle assemblies functionalized with biomolecules can be used as a molecular-scale ruler that operates over a length scale much larger than that in the fluorescence energy transfer metrology that is routinely used in biology. Practical applications of this concept in systems biology, such as imaging of the motion of molecular motors or of structural changes in proteins and in DNA (figure 28).

A final example for a potentially far-reaching application is the use of particles composed of a dielectric core and a metallic shell in cancer treatment. These particles, when injected into the human body, are selectively bound to malicious cells, whereupon laser irradiation at a precisely engineered plasmon resonance wavelength is used to heat the particles and thereby destroy the cells.

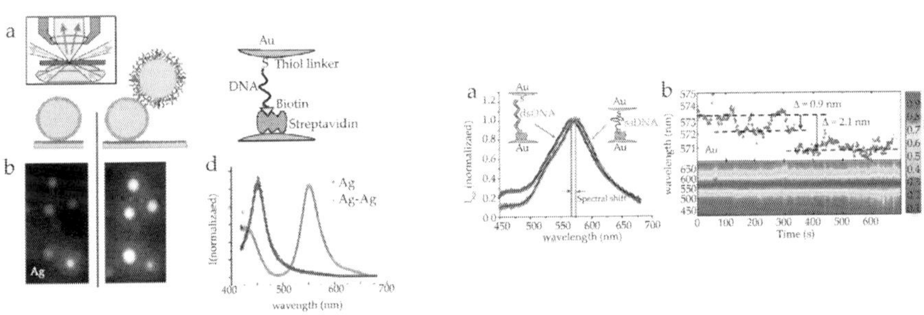


**Figure 28:** Single functionalized Ag nanoparticles as anchors for DNA molecules. When attaching a DNA molecule to two Au particles, the shift in resonance can be used to measure the distance between them (change from blue to blue-green). This allowed to study the dynamics of structural changes of the DNA molecule.

## REFERENCES

1. Anger, Pascal, Bharadwaj, Palash, and Novotny, Lukas (2006). Enhancement and quenching of single-molecule fluorescence. Phys. Rev. Lett., 96:113002.
2. Avrutsky, Ivan (2004). Surface plasmons at nanoscale relief gratings between a metal and a dielectric medium with optical gain. Phys. Rev. B, 70:155416.
3. Babadjanyan, A. J., Margaryan, N. L., and Nerkarayan, Kh. V. (2000). Superfocusing of surface polaritons in the conical structure. J. Appl. Phys., 87(8):3785–3788.
4. Bai, Benfeng, Li, Lifeng, and Zeng, Lijiang (2005). Experimental verification of enhanced transmission through two-dimensionally corrugated metallic films without holes. Opt. Lett., 30(18):2360–2362.
5. Baida, F. I. and van Labeke, D. (2002). Light transmission by subwavelength annular aperture arrays in metallic films. Opt. Commun., 209:17–22.
6. Bakker, Reuben M., Drachev, Vladimir P., Yuan, Hsiao-Kuan, and Shalaev, Vladimir M. (2004). Enhance transmission in near-field imaging of layered plasmonic structures. Opt. Express, 12(16):3701–3706.
7. Barnes, W. L., Murray, W. A., Dintinger, J., Devaux, E., and Ebbesen, T. W. (2004). Surface plasmon polaritons and their role in the enhanced transmission of light through periodic arrays of subwavelength holes in a metal film. Phys. Rev. Lett., 92(10):107401.
8. Cao, Hua, Agrawal, Amit, and Nahata, Ajay (2005). Controlling the transmission resonance lineshape of a single subwavelength aperture. Opt. Express, 13(3):763–769.
9. Chan, H. B., Marcet, Z., Woo, Kwangje, Tanner, D. B., Carr, D. W., Bower, J. E., Cirelli, R. A., Ferry, E., Klemens, F., Miner, J., Pai, C. S., and Taylor, J. A. (2006). Optical transmission through double-layer metallic subwavelength slit arrays. Opt. Lett., 31(4):516–518.

10. Chang, Shih-Hui, Gray, Stephen K., and Schatz, George C. (2005). Surface plasmon generation and light transmission by isolated nanoholes and arrays of nanoholes in thin metal films. Opt. Express, 13(8):3150–3165.
11. Charbonneau, Robert, Berini, Pierre, Berolo, Ezio, and Lisicka-Shrzek, Ewa (2000). Experimental observation of plasmon-polariton waves supported by a thin metal film of finite width. Opt. Lett., 25(11):844.
12. Daniels, Jacquitta K. and Chumanov, George (2005). Nanoparticle-mirror sandwich substrates for surface-enhanced Raman scattering. J. Phys. Chem. B, 109:17936–17942.
13. de Abajo, F. J. Garcia (2002). Light transmission through a single cylindrical hole in a metallic film. Opt. Express, 10(25):1475–1484.
14. El-Sayed, Ivan H., Huang, Xiaohua, and El-Sayed, Mostafa A. (2005). Surface plasmon resonance scattering and absorption of anti-EGFR antibody conjugated gold nanoparticles in cancer diagnostics: Applications in oral cancer. Nano Letters, 5(5):829–834.
15. Fang, Nicolas, Lee, Hyesog, Sun, Cheng, and Zhang, Xiang (2005). Sub-diffraction-limeted optical imaging with a silver superlens. Science, 308:534–537.
16. Farrer, Richard A., Butterfield, Francis L., Chen, Vincent W., and Fourkas, John T. (2005). Highly efficient multiphoton-absorption-induced luminescence from gold nanoparticles. Nano Lett., 5(6):1139–1141.
17. Félidj, N., Aubard, J., Lévi, G., Krenn, J. R., Schider, G., Leitner, A., and Aussenegg, F. R. (2002). Enhanced substrate-induced coupling in two-dimensional gold nanoparticle arrays. Phys. Rev. B, 66:245407.
18. Ganeev, R. A., Baba, M., Ryasnyansky, A. I., Suzuki, M., and Kuroda, H. (2004). Characterization of optical and nonlinear optical properties of silver nanoparticles prepared by laser ablation in various liquids. Opt. Commun., 240:437–448.

19. García-Vidal, F. J., Lezec, H. J., Ebbesen, T. W., and Martín-Moreno, L. (2003a). Multiple paths to enhance optical transmission through a single subwavelength slit. Phys. Rev. Lett., 90(21):213901.
20. García-Vidal, F. J., Martín-Moreno, L., Lezec, H. J., and Ebbesen, T. W. (2003b). Focusing light with a single subwavelength aperture flanked by surface corrugations. Appl. Phys. Lett., 83(22):4500–4502.
21. García-Vidal, F. J., Martín-Moreno, L., and Pendry, J. B. (2005a). Surfaces with holes in them: new plasmonic metamaterials. J. Opt. A: Pure Appl. Opt., 7:S97–S101.
22. Haes, Amanda J., Hall, W. Paige, Chang, Lei, Klein, William L., and van Dyne, Richard P. (2004). A localized surface plasmon resonance biosensor: First steps toward and assay for Alzheimer's disease. Nano Letters, 4(6):1029–1034.
23. Hartschuh, Achim, Sánchez, Erik J., Xie, X. Sunney, and Novotny, Lukas (2003). High-resolution near-field Raman microscopy of single-walled carbon nanotubes. Phys. Rev. Lett., 90(9):095503.
24. Hayazawa, Norihiko, Saito, Yuika, and Kawata, Satoshi (2004). Detection and characterization of longitudinal field for tip-enhanced Raman spectroscopy. Appl. Phys. Lett., 85(25):6239– 6241.
25. Hibbins, Alastair P., Lockyear, Matthew J., Hooper, Ian R., and Sambles, J. Roy (2006). Waveguide arrays as plasmonic metamaterials: transmission below cutoff. Phys. Rev. Lett., 96:073904.
26. Imura, Kohei, Nagahara, Tetsuhiko, and Okamoto, Hiromi (2005). Near-field optical imaging of plasmon modes in gold nanorods. J. Chem. Phys., 122:154701.
27. Jette-Charbonneau, Stephanie, Charbonneau, Robert, Lahoud, Nancy, Mattiussi, Greg, and Berrini, Pierre (2005). Demonstration of Bragg gratings based on long-ranging surface plasmon polariton waveguides. Opt. Express, 13(12):4674–4682.

CHAPTER 2

# LIGHT SCATTERING BY SMALL METALLIC PARTICLES

## INTRODUCTION

The production, control, manipulation and use of light are at the core of many technologies. Light scattering plays key roles in all of these. Of course, the scattering of light by small particles has a long history, where it was studied in contexts such as cumulus clouds, the color of the sky and rainbows, and used in various glass artifacts and windows from the middle ages. The remarkable fact is that such a classical topic is the basis of many fundamentally new and unexpected scientific and technological advances. The key is the current focus on the nanoscale and especially near-field effects at the nanoscale, while much of the older classical study was oriented towards the accessible far-field behavior.

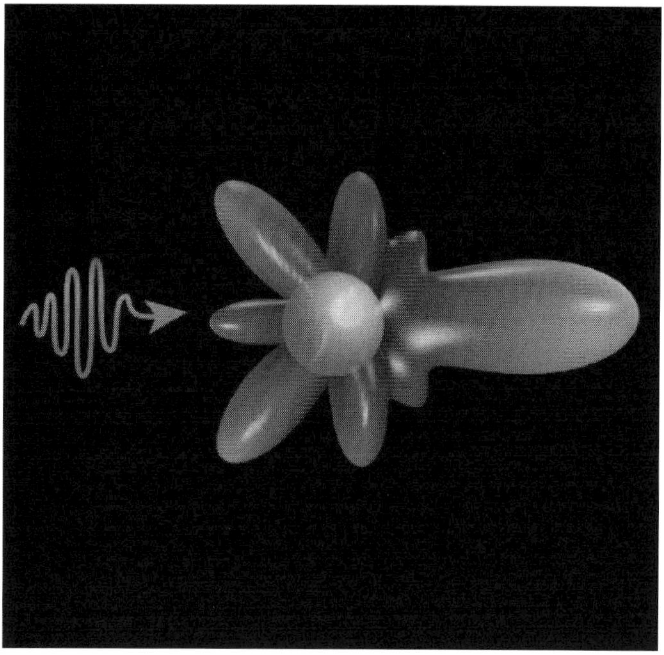

## 2.1 LIGHT SCATTERING AND SURFACE PLASMON'S

The production, control, manipulation and use of light are at the core of many technologies. Light scattering plays key roles in all of these. Of course, the scattering of light by small particles has a long history, where it was studied in contexts such as cumulus clouds, the color of the sky and rainbows, and used in various glass artifacts and windows from the middle ages. The remarkable fact is that such a classical topic is the basis of many fundamentally new and unexpected scientific and technological advances. The key is the current focus on the nanoscale and especially near-field effects at the nanoscale, while much of the older classical study was oriented towards the accessible far-field behavior. More specifically, there have been fascinating developments in regard to the light scattering by nanosized particles, including metal particles and surfaces, where localized surface plasmons can be excited leading to optical resonance phenomena. Small particles with surface plasmons can be used to detect the fluorescence of single molecules, enhance Raman scattering, resonantly transfer energy of excitons and create nanosized quantum amplifiers of optical energy. Potential practical uses include small-scale sensing techniques, numerous biomedical applications, manipulation of light for solar energy technologies and others.

Here we provide a short review emphasizing the nano-optics of small particles, near-field effects and the fundamental theoretical basis for their description. We begin with a review of the classical light scattering theory for spherical particles based on the quasistatic (Rayleigh) approximation and the general Mie theory. Scattering by dielectric particles is discussed along with the new topic of optical trapping. We discuss plasmon resonances and light scattering on small metallic particles, which is a subject that has been renewed by a series of new findings, including anomalous scattering with an inverted hierarchy of resonances and Fano resonances. The breakdown of the general Drude model for dielectric function at very small particle sizes and the resulting effects are discussed. Finally, we review the stimulated radiation from the surface plasmons of small particles along with concepts for new kinds of lasers based on nanolasing related to surface plasmons coupled to an active medium (so-called spasers).

We start with a summary of the basic concepts that remain useful in understanding light scattering, focusing on the case of spherical particles. Light scattering by small particles is one of fundamental problems of electrodynamics. As mentioned, it is a classical subject for which theory was developed long ago. This theory includes both the near and far field description. However, until recently the near field was inaccessible to experiments, and the interest was focused on far field effects. Now with advances in nanotechnology and nano-optics, the richness of the near field theory is being exploited. This includes the production by scattering of very high light intensities with spatial variations shorter than the wavelength—a phenomenon that enables rich new physics, both linear and nonlinear, at the nanoscale.

The physical understanding of light scattering by small particles began with the electric dipole concept, introduced by Lord Rayleigh in 1871. One starts with the assumption that the electromagnetic phase is constant over the region of interest, which is natural since the size of small particle considered is less than the wavelength of light. Then the homogeneous field of the incident light induces a polarization, which in turn results in light scattering. Higher order scattering modes, such as quadrupole and octupole, are not considered at this level. The polarization (i.e. the induced dipole) of materials in response to electromagnetic fields is determined by the dielectric function.

The dielectric function of a material (at energies above the phonon energies) is determined by its electronic structure. It is practical to calculate dielectric functions from first-principles band theory and often good agreement with experiment is found. However, for analyzing optical properties, it is often useful to approximate the optical properties of solids using the classical harmonic oscillator formalism introduced by Lorentz. In the Lorentz model, the dielectric function of non-conducting materials can be expressed as

$$\varepsilon = 1 + \frac{f}{\omega_0^2 - \omega^2 - i\gamma\omega}$$

where $f$ and $\omega_0$ are a phenomenological oscillator strength and frequency representing the bound electrons and $\gamma$ is a damping constant. This naturally leads to the Sellmeier formula for the refractive index, $1/(n^2-1)=-A/\lambda^2+B$, where $n$ is the refractive index, $\lambda$ is the wavelength and $A$ and $B$ are material-dependent quantities. This formula and generalizations (e.g., to two or more oscillators) are very effective in fitting the optical constants of real materials.

Many physical phenomena can be very simply understood even in the simplest one oscillator theory. For example, the dispersion of light by prisms or water drops is explained by the frequency dependence of the refractive index. This follows the normal dispersive behavior (refractive index increases with energy) for materials like glass and water. This originates in the fact that the energy ($\omega_0$) of the effective oscillator for transparent materials such as these is generally much larger than the frequency of visible light. For metals, the contributions of free electrons need to be added. This yields a model known as the Lorentz–Drude model. This model has

$$\varepsilon = 1 - \frac{\omega_p^2}{\omega^2 + i\gamma_e\omega} + \sum_j \frac{f_j}{\omega_j^2 - \omega^2 - i\gamma_j\omega}$$

where the sum on $j$ is over different oscillators. The free electron part is due to the electron plasma of the metal, which is described by the parameters $\omega_p$ and $\gamma_e$ which represent the resonant frequency and damping constant of bulk plasma.

It has long been recognized that all linear optical phenomena can in principle be modeled by solving the Maxwell's equations with known dielectric functions of the media. This understanding, while correct, by itself yielded relatively little direct physical insight into light scattering, at least prior to the development of modern computers and electromagnetic codes. This is because the vector electromagnetic equations resisted analytic solution, especially for complicated, but interesting, boundary conditions.

The earlier work on light scattering by small particles is mainly from Lorenz, Thomson and Clebsh. Actually, the exact solution has been obtained by Clebsh in 1861 in his paper 'Concerning reflection on a spherical surface' published in 1863, a year before Maxwell's work about electromagnetic theory of light. The breakthrough in understanding light scattering by spherical structures came from the work of Mie in 1908. He obtained a general rigorous solution, on basis of the electromagnetic theory, for the optical scattering by a homogeneous sphere with arbitrary size in a homogeneous medium, whatever the composition of the sphere and medium. The Mie theory solution is also applied directly to the scattering by any number of spheres if the distance between particles is large enough so that there are no coherent phase relations among the scattered light from different particles.

The general Mie theory of optical scattering is very useful in practice. Initially, interesting problems, such as the origin of rainbows and the solar corona, could be directly answered on basis of the Mie solution. So-called corrected Mie theory gives the light scattering by structures with other regular shapes, such as ellipsoids with any size and cylinders with arbitrary radius. The basic optical properties of small particles made of different materials have been analyzed in detail *via* modifications of Mie theory, such as the Gans modification (for spheroidal particles, e.g., plasmonic gold and silver nanoparticles with elongated shapes) and the Maxwell–Garnett equations (providing an effective medium approach). Of course, there are also interesting questions related to light scattering by particles that cannot be directly described with Mie theory. An example is the electromagnetic hot spot between two nearby particles, which depends on coherency.

As mentioned, it is remarkable that even though more than 100 years have passed since the introduction of Mie's general theory of light scattering by a sphere, new and exciting physics associated with light scattering by small particles continues to be found. Examples include the finding of giant optical resonances with an inverted hierarchy (e.g., the quadrupole resonance is more intense than the dipole) in scattering by small particles with negative dielectric susceptibility and weak dissipation, and anomalous scattering with the complicated near-field structures, such as the vortices, unusual frequency dependence, *etc.* In addition, the Fano resonances, which are well known in quantum physics, were discovered in optics of small metallic particles. Turning to nanotechnology and nano-optics, there is an increasing focus on control of optical energy in subwavelength structures—a field that is leading to many new ideas and remarkable experiment results. These include nanolenses and nano-antennas. The localized surface plasmons in spherical core–shell structures can result in so-called spasers, with resonantly coupled transitions with excitons from dye molecules in the shell layer. The coupling between localized surface plasmons from closely separated small particles results in electromagnetic hot spots. Huge light scattering is found on anisotropic spherical particles with an active mechanism. These remarkable results based on light scattering by small particles suggest many potential applications, including solar energy technologies and nanoscale lasers.

## 2.2 LIGHT SCATTERING BY SPHERICAL PARTICLES IN THE QUASISTATIC APPROXIMATION AND BEYOND

In the quasistatic approximation, light scattering by sphere particles is addressed by solving the Laplace equation for the scalar electric potential

$$\nabla^2 \Phi = 0 \text{ and } E = -\nabla \Phi \tag{1}$$

with continuous boundary conditions

$$\Phi_1 \big|_a = \Phi_2 \big|_a \text{ and } \varepsilon_p \frac{\partial \Phi_1}{\partial r}\bigg|_a = \varepsilon_m \frac{\partial \Phi_2}{\partial r}\bigg|_a \tag{2}$$

Where $E = E_0 \vec{z}$, $a$, $\varepsilon_p$ and $\varepsilon_m$ are the electric field along $z$ direction, radius of particle, dielectric function of the particle and that of the medium, respectively. By comparing the scattering potential from the Laplace equation with that of a dipole, the effective dipole moment can be expressed as

$$P = 4\pi\varepsilon_m a^3 \frac{\varepsilon_p - \varepsilon_m}{\varepsilon_p + 2\varepsilon_m} E_0$$

The cross-sections for scattering ($C_{sc}$) and adsorption ($C_{abs}$) are obtained from the scattering field radiated by this dipole, which is induced by the incident plane wave (i.e., the dipole moment $P$). The resulting expressions are

$$C_{sc} = \sigma_{geom} Q_{sc}, \quad Q_{sc} = \frac{8}{3} q^4 \left| \frac{\varepsilon_d - 1}{\varepsilon_d + 2} \right|^2 \text{ and}$$

$$C_{abs} = \sigma_{geom} Q_{abs}, \quad Q_{abs} = 4q\mathrm{Im}\left[\frac{\varepsilon_d - 1}{\varepsilon_d + 2}\right] \tag{3}$$

where $\sigma_{geom}=\pi a^2$ is the geometrical cross-section. We introduce the dimensionless cross-sections of scattering ($Q_{sc}$) and adsorption ($Q_{abs}$), the dimensionless size $q=ka$ and the relative dielectric function $\varepsilon_d=\varepsilon_p/\varepsilon_m$ for convenience in the discussion that follows. In the formula, $k$ is the wave vector in medium.

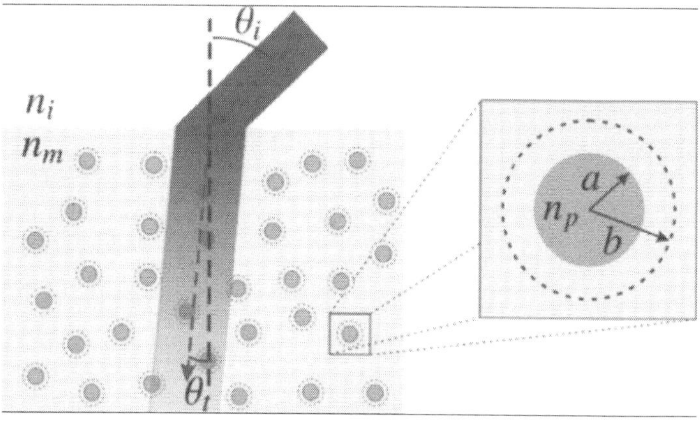


The above formula implies that as the particle size is decreased, the efficiency of adsorption will dominate over the scattering efficiency. Therefore, particles with very small size will be difficult to detect by light scattering. In addition, one may note that there is a resonant enhancement for scattering and adsorption when the condition $\mathrm{Re}(\varepsilon_d)=-2$ (the Fröhlich condition) is satisfied. This resonance is due to resonant excitation of the dipole surface plasmon. With the Drude model of dielectric function, the frequency of the dipole surface plasmon can be expressed as $\omega_{sp} \approx \omega_p/\sqrt{3}$, if the frequency of electron collisions $\gamma$ is small. The spectrum for the scattering cross-section then has a Lorentz line shape,

$$Q_{sc}^{(Ra)} = \frac{8}{3} \frac{\omega_{sp}^4}{\left(\omega^2 - \omega_{sp}^2\right)^2 + \omega^2 \gamma^2} q^4 \tag{4}$$

The above theory, however, does not capture the size effects on the spectra, including the changes of position and width of plasmon peak. Retardation effects in larger particles result in breakdown of the quasistatic approximation.

Within the Mie solution, the dimensionless cross-sections for scattering, extinction and absorption can be expressed as

$$Q_{sc}^{Mie} = \frac{2}{q^2} \sum_{l=1}^{\infty} (2l+1)\left\{|a_l|^2 + |b_l|^2\right\},$$

$$Q_{ext}^{Mie} = \frac{2}{q^2} \sum_{l=1}^{\infty} (2l+1)\mathrm{Re}(a_l + b_l) \text{ and } Q_{abs}^{Mie} = Q_{ext}^{Mie} - Q_{sc}^{Mie} \tag{5}$$

The scattering amplitudes $a_l$ and $b_l$ are defined as follows,

$$a_l = \frac{F_a^e(l)}{F_a^e(l) + iG_a^e(l)} \text{ and } b_l = \frac{F_b^m(l)}{F_b^m(l) + iG_b^m(l)} \tag{6}$$

Where $F_a^e(l)$, $G_a^e(l)$, $F_b^m(l)$ and $G_b^m(l)$ are related to the Bessel and Neumann functions. These general expressions are rather cumbersome to present here and can be found in literature. Expanding the Bessel and Neumann functions in power series for small $q$, the functions $F_a^e(l)$, $G_a^e(l)$, $F_b^m(l)$ and $G_b^m(l)$ can be expressed as,

$$F_a^e(l) \approx q^{2l+1} \frac{(l+1)}{[(2l+1)!!]^2} \tilde{n}^l \left(\tilde{n}^2 - 1\right)$$

$$G_a^e(l) \approx \tilde{n}^l \frac{l}{2l+1} \left\{\tilde{n}^2 + \frac{l+1}{l} - \frac{q^2}{2}\left(\tilde{n}^2 - 1\right)\left[\frac{\tilde{n}^2}{2l+3} + \frac{l+1}{l(2l-1)}\right]\right\}$$

$$F_b^m(l) \approx -\frac{\tilde{n}q^2}{2l+1} F_a^e(l)$$

$$G_b^{m}(l) \approx -\tilde{n}^{l+1}\left[1+\frac{1-\tilde{n}^2}{2(2l+1)}q^2\right] \tag{7}$$

Where $\tilde{n}=\sqrt{\varepsilon_d}=n_d+i\kappa_d$ is the relative complex refractive index with the relative real refractive index $n_d$ and the relative absorption index $\kappa_d$. Near the dipole scattering resonance, only the term $l$=1 in Mie formula needs to be considered. For small particles, the magnetic scattering amplitudes $b_l$ can be ignored. Then the scattering cross-section becomes $Q_{sc}^{Mie} \approx 6|a_1|^2/q$, where $a_1 = a_l(l=1)$, is determined by Equation (6). At the Fröhlich resonance condition in the formula for $F_a^e(l)$ and $G_a^e(l)$, the scattering cross-section due to the dipole resonance can be expressed as,

$$Q_{sc}^{dip-Mie} \approx \frac{8}{3}\frac{\omega_{sp}^4}{\left(\omega^2-\omega_{sp}^2\right)^2+{}^4/_9 q^6\omega_{sp}^4}q^4 \tag{8}$$

One may note that the $\gamma$ parameter (corresponding to dissipative losses due to electron collisions) is ignored in the derivation. The spectrum is found to have a Lorentz profile with an effective parameter $\gamma_{eff}$, which has a similar role as the dissipation parameter in the Rayleigh spectrum and can be expressed as $\gamma_{eff}={}^2/_3\omega_{sp}q^3$

This damping is due to the radiative losses of plasmons. This is different from the dissipative loss term ($\gamma$) in Rayleigh formula. Therefore, the singularity in the corrected scattering formula is removed due to radiative losses, even if dissipative losses are neglected.

Intuitively, the red shift of the peak in the dipole resonance with increasing size is due to the weakening of the restoring force. This is because the distance between charges on opposite sides of the particle increases with size and so their interaction decreases. The red shift of resonance can be addressed directly by numerical solution of the Mie theory equations. The scattering cross-section, $C_{sc}$ from the dipole term neglecting dissipative losses is plotted in Figure 1a. The red shift is clearly seen. One may also note that the resonant scattering cross-section is similar for the different sizes. However, with dissipative losses, the scattering cross-section falls quickly with decreasing particle size (Figure 1b). Clearly, the effect of dissipative losses on the scattering increases with decreasing size. The effect of the dielectric function of the medium $\varepsilon_m$ on the plasmon resonance of small metal particles is contained in the parameters $q$ and $\varepsilon_d$. From Equation (3), $C_{abs}$ is proportional to

$$\frac{\varepsilon_m^{3/2}\mathrm{Im}(\varepsilon_p)}{[\mathrm{Re}(\varepsilon_p)+2\varepsilon_m]^2+[\mathrm{Im}(\varepsilon_p)]^2}$$

The Fröhlich resonance condition then requires a decrease in $\mathrm{Re}(\varepsilon_p)$, for an increase in $\varepsilon_m$. For most metals, such as Au and Ag, $\mathrm{Re}(\varepsilon_p)$ decreases and $\mathrm{Im}(\varepsilon_p)$ increases with decreasing frequency around the resonance. Therefore, the peaks in the absorption spectra are shifted to longer wavelengths and become broader and more intense with the increase in $\varepsilon_m$.

## 2.3 OPTICAL SCATTERING BY DIELECTRIC PARTICLES

The light scattering by dielectric particles is easy to calculate using the Mie theory. Figure 2a and 2b shows two typical scattering curves for particles of refractive index $n$=1.33 and $n$=1.97 (that of water and glass).

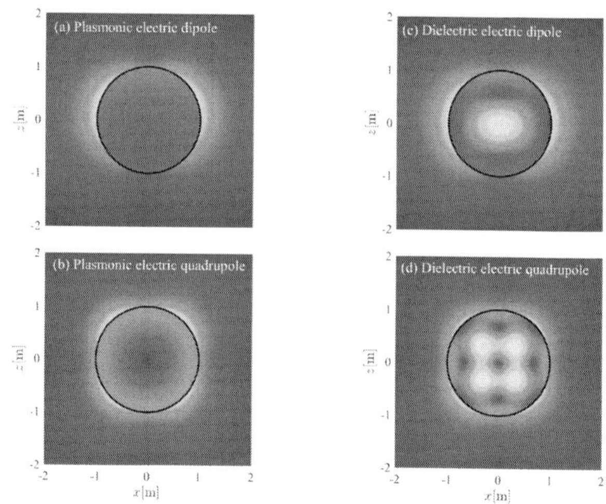


The curves have a series of maxima and minima. Other dielectric spheres with different refractive indices show similar behavior. Finally, we note that in the limit of an extremely large particle the scattering cross-section is twice as large as the geometrical cross-section. All these results for the scattering and extinction cross-sections were given 50 years ago. These results, taken in the far field, are connected with the study of light transmission in mist, fogs, and cloud chambers and so on, but the near field regions for dielectric spheres have received attention much more recently.

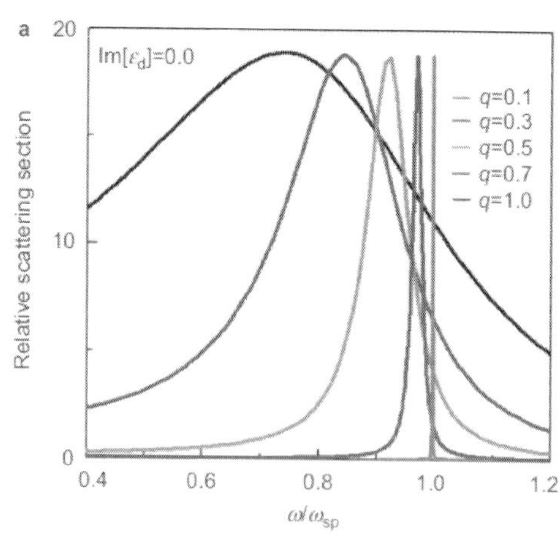

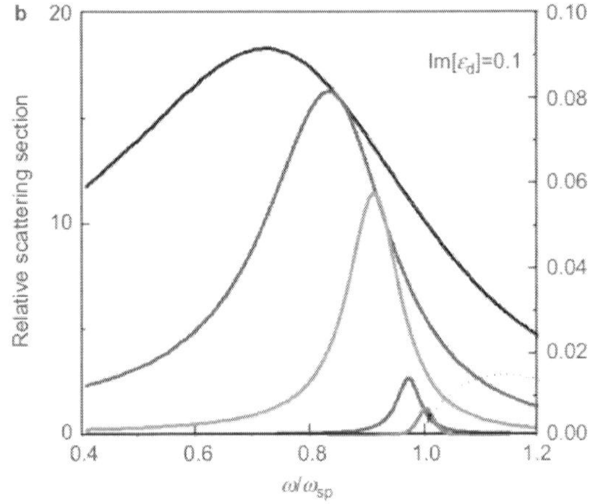


**Figure 1:** Illustration of the red shift of the dipole resonance, showing the relative scattering cross-section as a function of frequency for the different size parameters q in the cases of (a) Im[e]50 and (b) Im[e]0.1.

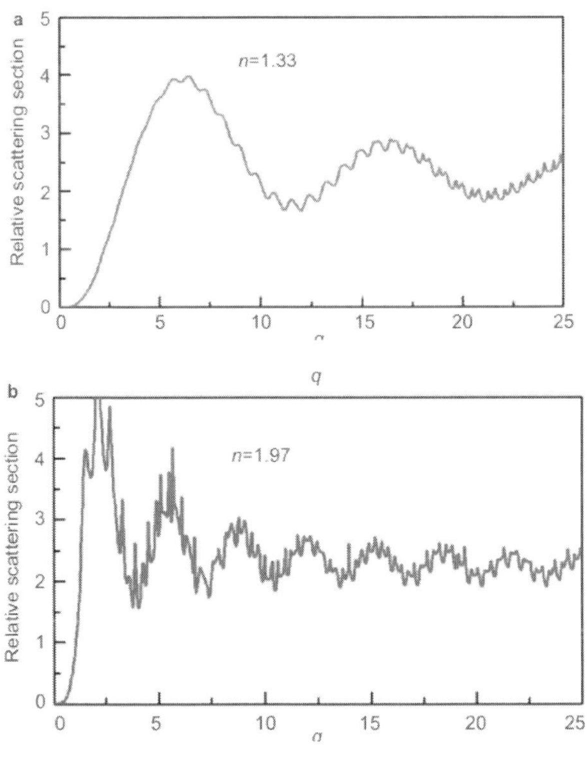

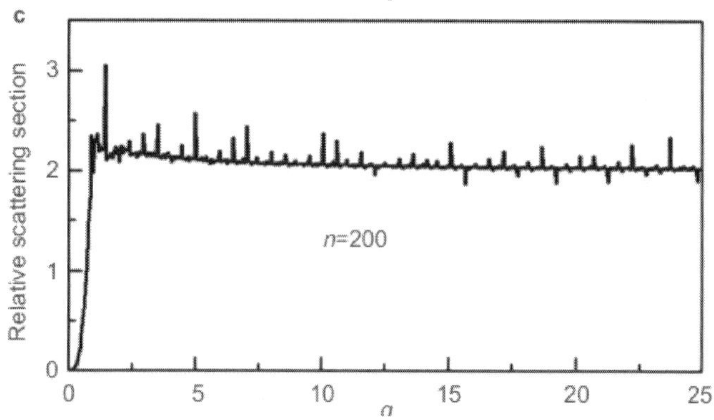


**Figure 2**: Relative scattering cross-sections of (**a**) a water particle with reflective index $n$=1.33, (**b**) a glass particle with reflective index $n$=1.97 and (**c**) ideal particle with very high reflective index $n$=200 as the functions of dimensionless size parameter $q$. The first maximum of $Q_{sc}$ happens for the quantity $2q(n-1)\sim4$. The case of refractive index $n$=200 represents $n\sim\infty$. $Q_{sc}$ approaches $2\pi a^2$ at large $q$.

For small size dielectric particles, the scattering cross-section will increase with increasing refractive index. This result follows from the Rayleigh theory and is indicative of the important role of dipole scattering. The squares of the intensity of the electric fields in the near field for different refractive indices are shown in Figure 3a–3c. The electric field intensity increases in the near field region with increasing refractive index. However, the near field electric field intensity from the dipolar mode does not become stronger as the size of dielectric particle decreases. Instead, the maximal value of electric field intensity increases along with a change in the spatial configuration, when the particle size increases in Figure 3d–3f.

Usually, Mie scattering will dominate when the particle size is larger than a wavelength. As shown in Figure 3d–3f, Mie scattering produces complex field distribution patterns reminiscent of directional antennas. This effect is sometimes referred to as a photonic nanojet. From the point of view of geometrical optics, the increase in the scattering field intensity with a more intense forward lobe reflects the fact that a dielectric sphere with large size behaves as a convex lens.

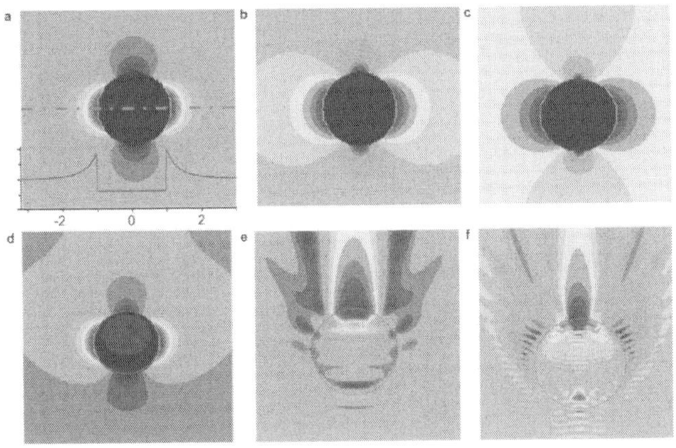


**Figure 3:** Near-field distribution of the square of electric field intensity for different dielectric particles including the small particles with the radius $R$=1.8 nm and different reflective indices (**a**) $n$=1.33, (**b**) $n$=2.5 and (**c**) $n$=4 (scattering light $\lambda$=496 nm) and with the reflective indices $n$=1.33 and (**d**) different radii $R$=50 nm, (**e**) $R$=500 nm and (**f**) $R$=1000 nm (scattering light $\lambda$=500 nm). In **a–f**, $E_{max}^2=1.90E_0^2$, $5.15E_0^2$, $7.10E_0^2$, $2.20E_0^2$, $16.0E_0^2$ and $38.0E_0^2$, respectively.

Localized regions of high intensity, such as those near the particle, can trap dielectric particles due to gradient forces (these are forces that arise because a particle with a dielectric constant higher than medium will lower the energy if it moves to the location of the maximum electric field intensity). This effect was observed by Ashkin *et al*. more than 30 years ago.

The related optical tweezers technique, in which light is used to manipulate small particles, has been applied in many different fields, especially in medicine, biology and biophysics, where biologically inert particles can be functionalized and then manipulated using light. Specifically, small dielectric particles can be picked up and controlled by tightly focused visible light lasers. It is interesting that this qualitative effect can also be understood just with ray optics.

A photon with energy $\hbar\omega$ carries the momentum $\hbar k$ If the photon is adsorbed by an object, a force $F$ on the object due to the transfer of momentum will be produced and given by the formula $F=n_m P/c$, where $n_m$ is the refractive index of the surrounding medium, $P$ is the power of light beam and $c$ is the light speed in vacuum. A dimensionless quantity $Q_{et}$ which is used to describe the efficiency of trapping light by a particular object with any shape, is defined by the formula $Q_{et}=n_m P/cF$. The refraction of light by a transparent object will result in the reaction force acting on the object, since the momenta of the photons are changed. For the plane perpendicular to the direction of light beam, an intensity profile with high symmetry will result in a force that will tend to move the object into the center of beam, since the force from the refraction, which points to the center, will provide a restoring force when the position of the object deviates from the center, shown in Figure 4. If the focus of the beam just is above the object, a force will be generated to lift the object up towards the focus.

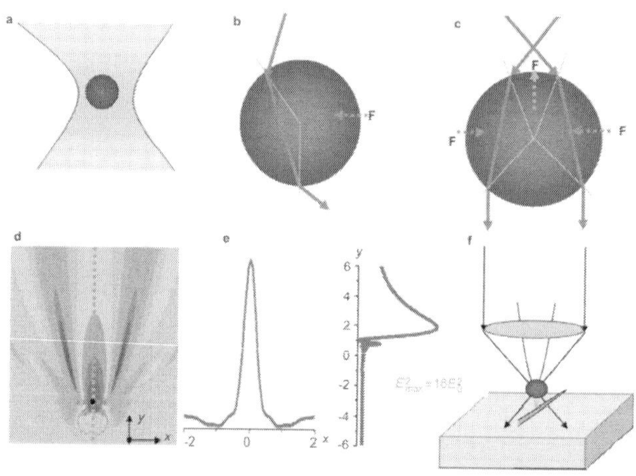

**Figure 4**: Schematic presentation of an optical tweezer (**a**), optical force from the view point of optical rays (**b, c**), using the combination of optical trapping and nanojets for optically scanning samples on the surface of substrate (**f**) and the near-field distribution of the square of electric field density of a glass sphere ($n$=1.5) in water ($n$=1.33) with radius $R$=800 nm scattering light $\lambda$=500 nm (**d**) and the distribution of $|E|^2$ along the dotted lines in (**d**) with $E_{max}^2=16E_0^2$ (**e**).

The concepts of geometrical optics are simple and intuitively appealing, but they are not strictly applicable to the case of particles that have sizes below the wavelength. When one considers the electric field in the near-field region, one sees that the force can be separated into two parts (note that both force terms necessarily imply a rate of momentum change of the light field). One is the force due to the intensity gradient and another one is from the scattering of light. The scattering force from the light beam with the intensity $I_0$ can be expressed as $F_{SC}=I_0 C_{SC} n_m/c$. The gradient force due to the intensity gradient can be given by the formula

$$F_{gra} = -n_m^3 a^3 \left(\frac{\varepsilon_d - 1}{\varepsilon_d + 2}\right) \nabla (E^2) / 2$$

where $E$ is the electric field near the particle. Clearly, the force is directed towards the higher intensity region and does not depend on the light propagation direction. The control of particles by optical trapping can be established when the gradient force exceeds the scattering force. In practice this is readily done for small particles, and precise three dimensional control of the particle position is possible by using optical interference or crossed beams to define a localized maximum in intensity.

An optically trapped particle in a viscose medium (viscosity $\eta$) will behave like a damped oscillator and thus the equation of motion will be, $m\ddot{x} + \beta\dot{x} + k_{st}x = 0$, where $k_{st}$ is the stiffness of optical trap and $\beta$ is the Stokes drag constant with the formula, $\beta=6\pi a\eta$. Further expressions for dielectric spheres including scattering can be obtained using the explicit partial-wave representations.

This type technique based on gradient forces has been broadly applied in very diverse areas. These range from disease diagnosis to gravitational detection. With new developments, optical tweezers can be used to detect biological compounds at the single-molecule level. Control of the motion of particles at nanometer scales with piconewton forces enables studies of molecular and nanoscale dynamics, for example the investigations of molecular motors. Furthermore, as mentioned and as shown in Figure 4, nanojets (directionally concentrated electromagnetic radiation) can be formed under large dielectric spheres. These nanojets are technologically important as they can be used to enhance Raman signals. Thus a dielectric particle, which is analogous to a nanosize convex lens, can induce a high light intensity under the particle (the nanojet), and this particle and its nanojet can be controllably moved and used as a nanoscale Raman probe using optical tweezers (Figure 4f).

## 2.4 LOCALIZED PLASMON'S OF METAL PARTICLES

It may be noted in the above discussion that the scattering efficiency from the dipole term in Mie theory increases as the particle size decreases in the small size region around the resonance frequency (see Equation (8)). This is clearly different from ordinary Rayleigh scattering. Ignoring dissipation, the high-order plasmon modes have resonant frequencies, $\omega_l^2 = \omega_p^2 l/(2l+1)$. Since all the amplitudes, $a_l$ tends to go to unity for the corresponding frequency, the scattering cross-sections of high-order plasmon modes can be expressed as

$$Q_{sc}(l) = \frac{2(2l+1)}{q(l)^2} = \frac{2(2l+1)^2}{l} \frac{c^2}{\omega_p^2 a^2}$$

Since the resonance frequencies of different modes are different and the resonant peaks of different modes are limited, the total scattering cross-section for each resonant frequency is given by $Q_{sc} \approx Q_{sc}(l)$. Therefore, anomalous scattering of light with an inverse hierarchy of the resonances will occur if the dissipation term in the dielectric function is very small, as shown in Figure 5. Usually, with the condition of the actual dissipation $\mathrm{Im}\,\varepsilon_d(\omega_l) << \frac{q^{2l+1}}{l[(2l-1)!!]^2}$, the anomalous scattering rises. Note, however, that ordinary Rayleigh scattering is restored when the size parameter $q$ tends to zero.

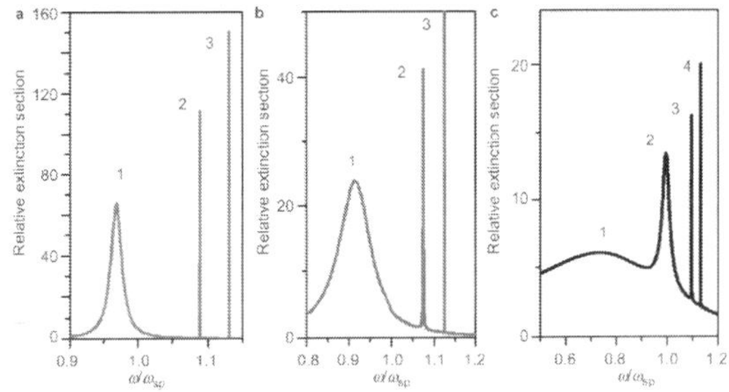


**Figure 5**: Relative extinction cross-section of a particle in the non-dissipative limit as a function of frequency for the different (**a**) $q$=0.3, (**b**) $q$=0.5 and (**c**) $q$=1.0. Note that the dielectric function is with the model $\varepsilon(\omega)=1-3(\omega_{sp}/\omega)^2$, where $\omega_{sp}$ is the resonance frequency of the dipole model in the limit of small $q$.

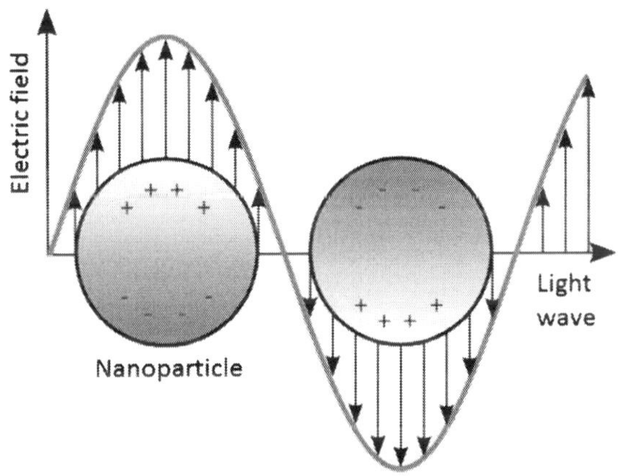


The Fano resonance found in 1961 is well known in quantum physics. Fano spectra arise from the constructive and destructive interference between a narrow resonant mode and a broad background spectral line. Fano spectra exhibit an asymmetric shape, specifically taking the form,

$$I(\omega)=\frac{(F\gamma+\omega-\omega_0)^2}{(\omega-\omega_0)^2+\gamma^2}$$

where $F$, $\omega_0$ and $\gamma$ are the Fano parameters, the position and the width of resonance, respectively. Fano resonances have been found in diverse quantum systems, such as quantum dots and tunnel junctions. Fano resonances are also expected to appear in light scattering. In plasmonic materials, the resonant peak of each plasmon mode has a very different linewidth. Therefore, different plasmon modes can coexist in the same frequency region. Then Fano resonances can arise due to the constructive and destructive interference of plasmon modes with different multipolarity. The resonant interference does not occur in the total optical cross-section, such as the scattering and extinction cross-section for a single particle. It is seen in differential scattering cross-sections, such as forward scattering (fs) and radar back scattering (rbs), with the formulas

$$Q_{\mathrm{fs}}^{\mathrm{Mie}} = \frac{1}{q^2}\left|\sum_{l=1}^{\infty}(2l+1)[a_l+b_l]\right|^2 \text{ and}$$

$$Q_{\mathrm{rbs}}^{\mathrm{Mie}} = \frac{1}{q^2}\left|\sum_{l=1}^{\infty}(2l+1)(-1)^l[a_l-b_l]\right|^2 \qquad (9)$$

Thelinewidthdecreasesquicklywithincreasingtheorderofplasmon mode, according to the formula, $\gamma_l = \dfrac{q^{2l+1}(l+1)}{[l(2l-1)!!]^2(d\varepsilon_d/d\omega)_l}$

Clearly, the interaction of a dipolar mode and a quadrupole mode is easiestduetotheradiativecoupling,especiallyfortherelativelysmall particles. The magnetic amplitude can be ignored, and then the low-energy interference of the electric dipole and quadrupole is given by formulas, $Q_{\mathrm{fs}} = (1/q^2)|3a_1+5a_2|^2$ and $Q_{\mathrm{rbs}} = (1/q^2)|3a_1-5a_2|^2$. $Q_{\mathrm{rbs}}$ and $Q_{\mathrm{fs}}$ as the functions of frequency are shown in Figure 6a, where a Fano resonance near the quadrupole resonance frequency is clearly seen.

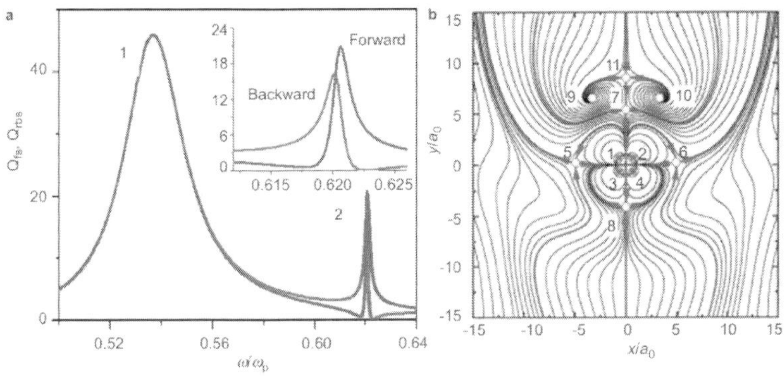


**Figure 6**: Fano resonance on Mie scattering by a small metal sphere (**a**) and energy flow in the quadrupole resonance with the singular vortices represented by the Poynting vector field (**b**). In **a**, Radar back scattering and forward scattering directions are indicated by the red line and blue line, respectively, the dielectric function $\varepsilon(\omega)$ is described by Drude model with the dissipation parameter $\gamma$=0.001$\omega_p$ and the particle size is $a$=0.8$c/\omega_p$. In **b**, $q$=0.3 and $\varepsilon_d$=−1.553, the blue line denotes the particle surface, the red lines indicate the separatrix. Figure reproduced with permission, **b** from Ref. ©2007 IOP.

The interference of incident and re-emitted light in the scattering process generates complex patterns in the near-field region. The energy flow, as represented by the Poynting vector, from the dipole has helicoidally shaped vortices, while that from the quadrupole is still more complex with vortices and singular points (Figure 6b). Higher-order modes can also interfere with the broad dipole mode as the size increases. However, it is important to note that the dissipative losses of plasmonic materials must be weak for the Fano resonance to appear, since the higher-order modes are rapidly suppressed when dissipative losses increase.

The Fano resonance of a single spherical particle is generally difficult to observe due to dissipative losses. If the widths and energy positions of plasmon modes can be modulated independently, the condition about the interference between a narrow discrete mode and a broad background resonance is easier to realize. An example is a non-concentric ring/disk cavity. The dipolar modes from disk and ring interact to result in a hybridized bonding mode and a broad higher-energy antibonding mode. The coupling between the quadrupolar mode from the ring and the antibonding dipolar mode due to the symmetry breaking of the non-concentric geometry can induce an enhanced Fano resonance. Related ideas can be also applied to the other plasmonic nanostructures, such as nanoshells, dolmen-type structures, and finite clusters of plasmonic nanoparticles and so on. In addition, Fano resonances frequently appear in photonic crystals, such as periodic metallic structures on a single-mode slab. The waveguide mode from the slab can couple with the plasmon modes of the metallic structures exited by the incident light. Optical Fano resonances have recently been found in electromagnetic metamaterials. The high asymmetrical profiles of Fano resonances suggests important applications, including novel sensors, as well as lasing and switching schemes.

In the case of non-magnetic particles, there are also other unconventional Fano resonances. An example is in the light scattering by small particles with large dielectric permittivity or with spatial dispersion. This kind of resonance in scattering by small particles is beyond the applicability of Rayleigh approximation. The electromagnetic modes excited by the incident wave which can interfere with each other have the same multipole moment $l$. This results in the conventional Fano resonances, while those modes with different $l$, which have spatial dispersion, can yield directional Fano resonances.

Fano resonances can also occur in light scattering by magnetic particles. This occurs with negative magnetic permeability ($\mu<0$) and positive dielectric permittivity ($\varepsilon>0$). In that case, the interference of different magnetic multipole modes can result in the Fano effect, such as that between magnetic dipole ($b_1$) and quadrupole ($b_2$). With the effective magnetic permeability, the effect of magnetic modes on the light scattering becomes important and the interference of electric and magnetic modes (Kerker effect) can occur. With the condition $\varepsilon=\mu$, the backward scattering gain is zero. It is also possible that the forward intensity is zero and the dipoles are out of phase, under the second Kerker condition. Excitingly, an unconventional forward–backward scattering asymmetry was recently observed experimentally in scattering by a single subwavelength sphere.

The Plasma is an important concept in physics. It is used to explain the energy losses of fast electrons in thin metal films. The theoretical work of Ritchie (1957) and the experimental work of Powell and Swan (1959) laid the groundwork for the study of surface plasmons by measurements of electron energy loss spectra. The optical properties of metallic materials in a low energy region are controlled mostly by collective plasmonic excitations of conduction electrons.

Surface plasmons can be excited by optical beams using a prism with the attenuated total reflection method as shown by Otto and Kretschmann *et al.* in 1968. Importantly, in the case of metallic particles, the finite surface area can localize the propagation of light and result in localized surface plasmons, which as mentioned have many current and potential applications.

The dielectric function of an ideal bulk metal at low energy can be phenomenologically expressed by the Drude model of free electrons. In modeling real metals, a term corresponding to Lorentz oscillators is usually introduced to describe the increase of the imaginary part of dielectric function $\mathrm{Im}(\varepsilon)$ due to the interband transitions. These can also be calculated in detail using first principles electronic structure methods. According to Fermi liquid theory, the conduction bands of metals are continuous near the Fermi surface and the low energy properties are like those of an electron gas, though renormalized from the free electron gas and with anisotropy and other complexities reflecting the crystal lattice and band formation. As such, given the band structure, the plasma frequency can be directly calculated from the band dispersion at the Fermi surface and important insights about the nature of the metallic state can be gained from its comparison with experiment. In non-cubic solids, the Drude plasma frequency has the form of a rank-2 tensor and therefore can be anisotropic. In any case, the presence of conduction electrons will result in intraband excitations within the conduction band by the creation of electron–hole pairs.

For noble metals, such as gold and silver, there are also interband transitions from lower-lying d-bands to the sp-hybridized conduction bands. These are the main causes of dissipative losses. In addition, there are other generally weaker processes including elastic and inelastic electron scattering, such as electron–electron, electron–phonon and electron–defect interactions. All these dissipative loss mechanisms can result in the non-radiative decay of plasmons and importantly can be phenomenologically described using the Lorentz–Drude dielectric model.

The dielectric functions of bulk gold and silver are shown in Figure 7e and 7g. Figure 7f and 7h shows the spatial configurations of the square of electric fields of gold and silver particles with radius $R$=1.6 nm at the dipole resonance Re($\varepsilon_d$)=–2. Clearly, the dissipative losses have an important effect on the intensity in the near field. The variation of the near-field electromagnetic intensity configurations with particle size is shown away from resonance in Figure 7a–7d. One may note that the strength of electric field in the near field does not increase with this parameter and that this is obviously different from the behavior illustrated for dielectric particles in Figure 3d–3f.

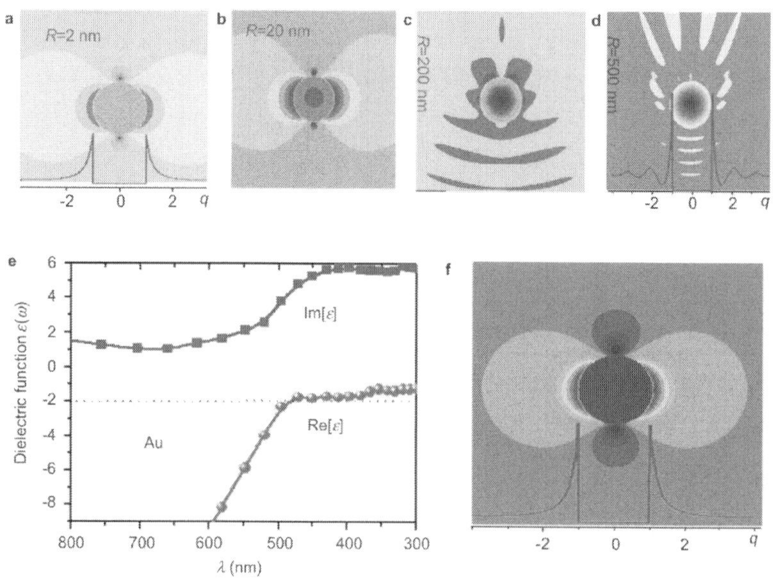

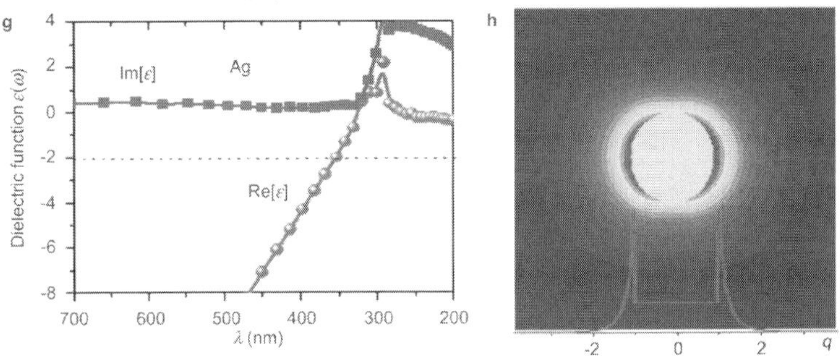


**Figure 7**: Near-field distribution of the square of electric field density on a silver particle scattering light $\lambda$=496 nm away from resonance with different sizes (**a–d**), where the dielectric constant is $\varepsilon$=−9.56+0.31i (the radius $R$=2 nm, $E_{max}^2$=14$E_0^2$; $R$=20 nm, $E_{max}^2$=16$E_0^2$; $R$=200 nm, $E_{max}^2$=17$E_0^2$; $R$–500 nm, $E_{max}^2$=16$E_0^2$), dielectric function and near-field distribution of electric field density on a small particle at the resonance condition for gold (**e**, **f**) and silver (**g**, **h**). In **f**, $\lambda$=481 nm, the radius of sphere is $R$=1.6 nm, dielectric constant is $\varepsilon$=−2.0+4.4i and $E_{max}^2$=11$E_0^2$. In **h**, $\lambda$=354 nm, the radius of sphere is $R$=1.6 nm, the dielectric constant is $\varepsilon$=−2.0+0.28i and $E_{max}^2$=457$E_0^2$.

The localized surface plasmon resonances of noble metal particles with the sizes of more than 10 nm have been well characterized experimentally. The understanding of plasmon resonances for smaller sizes is, however, still poor. This is because both experiment and theory are challenging for small particle sizes. In particular, both quantum effects and detailed surface interactions become important as the electrons interact more strongly with the surface including the spill-over of conduction electrons at the cluster surface, which complicates geometrical analysis. Quantitative predictions then require detailed calculations of the electronic structure for the actual atomic arrangements of the clusters of interest. For experiment, optical detection in the far field becomes difficult for small particles due to the size-dependent reduction in scattering intensity. Theoretically, time-dependent density functional theory-based methods are usually limited at present to particles with the sizes below 1–2 nm, but still useful insights have emerged. Methods that bring detailed quantum mechanical calculations to the longer length scales of interest would be very valuable in better understanding the size regime where quantum effects start to become important.

The first effect we mention is the red shift effect in the case of alkali–metal particles, which is due to the finite surface area. The red shift is understood in terms of the spill-over effect. At small size, the electronic density profile will extend beyond the nominal surface. This is an effect of the high kinetic energy of the *s*-electrons that make up the conduction states of alkali metals. The resulting charge located outside the surface cannot be efficiently screened by the other electrons. So the polarizability is enhanced, which results in a decrease in the resonant frequency.

The effect of electron scattering at the surface may be described *via* a corrected dissipative loss term in the Drude model with the formula

$$\gamma' = \gamma_{\text{bulk}} + \frac{A v_F}{R}$$

where $\gamma_{\text{bulk}}$ is the parameter describing bulk dissipative losses, $R$ is the particle radius and $v_F$ is the Fermi velocity. $A$ is an empirical constant that can be set using fits of experimental data. This effect also results in a slight red shift of the resonant frequency.

Next we discuss the blue shift of the plasmon resonance of small non-alkali metal particles. This can be understood in terms *d*-electron contribution to the dielectric properties. In bulk materials, the Lorentz term in Lorentz–Drude model represents the contribution of interband transitions, involving the *s–d* interactions. The bulk plasmon resonant frequency is reduced from the unscreened value due to screening from *s–d* interactions. For example, the bare plasmon energy of Ag is reduced from 9.2 eV to 3.76 eV by screening. This is reduced at the surface of small particles as the *s* electrons spill out. The reduced screening that results will then yield a blue shift (the surface-to-bulk ratio increases as the size decreases). Detailed quantitative characterization of spill-over effects on the surface plasmons of small particles will depend on the development of atomistic methods for the surface electronic structure and excitations that can be applied for cluster sizes of interest. It may be that interesting new effects will emerge from studies that include detailed surface structures and interactions.

Both top-down and bottom-up methods have been adopted to analyze the size-dependent plasmon frequency. Starting with bottom-up approaches, cluster science has made an important contribution to the understanding of the optical properties of small particles both theoretically and experimentally. From the top-down, plasmon resonances can be studied by aberration-corrected transmission electron microscope imaging and monochromated scanning transmission electron microscope electron energy-loss spectroscopy. Microscopically, the free electron part of the Drude model can be modified to a phenomenological model of very small particles by considering the conduction electrons as an electron gas constrained in an infinite potential barriers. Then, the quantum size effects lead to a discrete set of energy levels near Fermi surface instead of a Fermi liquid. As discussed by Scholl *et al.*, these quantum size effects result in the blue shift of resonance frequency. This is in addition to the spill-out effect and the resulting weakened screening of *d* electrons, as discussed above. It is, however, important to note that there remain inconsistencies between experimental results from top-down and bottom-up approaches. Methods that can span the full size range of interest will be very helpful in developing a more quantitative understanding of the size dependence.

This is an exciting time for nanophotonic applications based on light scattering by particles. For applications, tuning of properties is important. One avenue for this is through the use of core-shell particles, including the special case of hollow core particles, instead of simple single component particles. For the spherical case, one has two dielectric functions, the core radius and the particle (core+shell) radius as parameters, instead of the single dielectric function and radius as tuning parameters for the single component case. One example of a core–shell particle used in light scattering is the case of metal particles in an aqueous solution. In this case, there may be chemical effects at the surface. In particular, the interface between particle and aqueous solution can be viewed as a double layer, and furthermore anodic or cathodic polarization can induce chemical changes due to anion adsorption or desorption, alloy formation and metal deposition including deposition of a shell with a different composition (e.g., Ag on Pd). Light scattering in such cases can be dealt with using core–shell models. Core–shell particles can be used to obtain new optical properties that single spherical particles do not exhibit. Furthermore, techniques for producing such particles are well developed.

The core–shell model has been studied using the full solution of Mie theory and also can be solved approximately using an electrostatic solution. The surface plasmon resonance condition becomes $\mathrm{Re}(\varepsilon_{sh}\varepsilon_{a}+\varepsilon_{m}\varepsilon_{b})=0$ with $\varepsilon_{a}=\varepsilon_{co}(3-2P_{ra})+2\varepsilon_{sh}P_{ra}$ and $\varepsilon_{b}=\varepsilon_{co}P_{ra}+\varepsilon_{sh}(3-P_{ra})$, where $\varepsilon_{co}$, $\varepsilon_{sh}$ and $\varepsilon_{m}$ are the dielectric functions of the core, the shell and the medium, respectively. The parameter $P_{ra}$ is the ratio of the shell volume to the total volume of particle. The result is that the plasmon resonance frequency depends on the ratio of the core radius to the total radius of particle.

Core–shell structures also introduce the important concept of plasmon hybridization. This provides a powerful principle for the design of complex metallic nanostructures. The plasmon modes of nanoshells (core–shell particles with an empty core, i.e. hollow shells) can be viewed as arising from hybridization of the plasmon modes of a nanoscale sphere and a cavity. This hybridization results in a low-energy bonding mode and high-energy antibonding mode, as mentioned in relation to the Fano effect. Many non-trivial nanostructures, such as gold nanostars and nanorice, have plasmons that can be understood in terms the interaction of the coupled plasmons of simpler systems.

The interparticle distance is another variable that can be used to produce new physics and applications. Examples are quantum tunneling and large electromagnetic enhancements at the junctions. The development of nanoscale fabrication methods has made possible the production of different forms of nanoparticle arrays. These include dimers, chains, clusters and uniform arrays. The simplest prototype, which can be used as a model, is a nanoparticle dimer. The interaction between localized plasmons and the interference of the electromagnetic fields from these plasmons are the two major factors that control the electromagnetic enhancements at the junctions. Different methods, such as the coupled dipole approximation, the finite difference time domain method and plasmon hybridization, have recently been used to understand the plasmonic properties of dimers. For the practical calculation, the temporal couple-mode model as an effective method is also developed. Within the framework of the hybridization concept, the dimer plasmons can be treated as bonding and antibonding combinations of the single particle plasmons. The shifts of the plasmons at large interparticle distance then follow the interaction between two classical dipoles, since this is the interaction that leads to the hybridization. At shorter distances, the plasmon shifts in dipolar models become stronger and vary more rapidly with distance. This is a consequence of hybridization (or mixing) coming from higher multipoles. In addition, new interesting effects beyond the hybridization models, such as Young's interference, have been recently observed in the plasmonic structures.

The plasmon modes for the symmetric nanoclusters can be analyzed based on plasmon hybridization with group theory. In addition, by introducing the symmetry breaking, nonsymmetrical nanoclusters can also be analyzed. In the case of uniform two-dimensional nanoparticle arrays with coupling to localized plasmons can result in a coherent interaction of the array with light propagating in the plane of the array. This results in a plasmonic band structure. In addition, in the subwavelength nanostructures, there is an substantial opportunity to obtain the superscattering if one can maximize the contributions from different channels. These can be enabling for a number of applications, including various photonic metamaterial applications and plasmonic lasers.

The polarization of a particle array can be expressed in the simple dipole approximation as $\alpha_{arr} = (1/\alpha - S)^{-1}$, where $\alpha$ and $S$ are the polarization of a single particle and the structure factor of array, respectively. There will be a geometric resonance when the wavelength of scattering light is commensurate with the periodicity of the particles array. The study on light scattering of uniform arrays of nanoparticles is strongly connected with the fields of photonic crystals and metamaterials. A detailed review was given by Garcia de Abajo, to which we refer the reader for details.

Finally, we note that nonlinear optical responses can be very strongly increased using nanoparticle plasmons. This is by two main mechanisms, namely through the field enhancement near the particle surface and *via* the sensitivity of resonance frequency to the dielectric function of surrounding medium. Some of the first work on nonlinear-optical effects of small metallic particles was on nanoparticle colloids. Extension of Maxwell–Garnett theory for the low-concentration limit of particles in the medium ($C_{ra} \ll 1$) can be used. The effective dielectric function of the nanoparticle colloids can be expressed as

$$\varepsilon_{eff} = \varepsilon_m + 3C_{ra}\varepsilon_m \frac{\varepsilon_p - \varepsilon_m}{\varepsilon_p + 2\varepsilon_m}$$

The resulting third-order susceptibility $\chi_m^{(3)}$ from plasmonic enhancement then can give rise to substantial optical Kerr effects. The formal electromagnetic description of small particle second-harmonic scattering (hyper-Raleigh, which should vanish in the dipole approximation due to inversion symmetry) was given by Dadap *et al.*, who described the second harmonic generation on a small centrosymmetric sphere on the basis of the Mie theory and determined the nonlinear susceptibilities and radiation pattern. This formalism, although based on the local bulk response, provides an approach for dealing with the contributions from non-local dipole and other multipole modes. Giant second-harmonic scattering has been observed in experiments on suspensions of small gold particles, and even for individual gold nanoparticles. Effective second-harmonic generation has also been studied in plasmonic structures with low symmetry, such as gold nanocones with a sharp tips, nano-apertures surrounded by gratings and non-centrosymmetric gold nanocups.

## 2.5 SPASERS: SURFACE PLASMON LASING FROM ACTIVE PARTICLES

In lasers, coherent electromagnetic energy is generally concentrated in the active region by a Fabry–Perot or similar resonator. Let us consider an ideal resonator with two perfect mirrors. The minimal distance between two mirrors for storing the electromagnetic field is half of the wavelength, as shown in Figure 8b. However, by using resonance between the electromagnetic field and a surface plasmon (described by a dipole model for a small metal particle), the electromagnetic energy can be concentrated in a much smaller region. The particle size for this can be estimated from the skin depth given by where $\lambda$ is the vacuum wavelength.

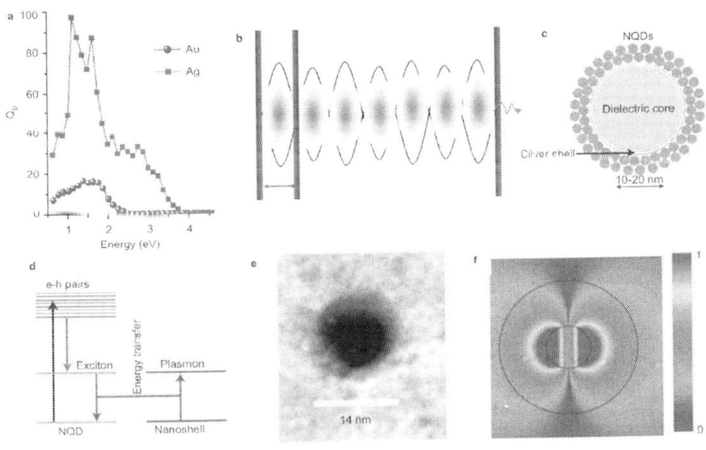


**Figure 8**: Quality factor of plasmons of gold and silver (**a**), schematic presentation of the electromagnetic wave in a Fabry–Perot resonator with the minimum possible distance between two mirrors (**b**) and the spasing mechanism (**c**–**f**). Schematic structure of a core–shell with a silver nanoshell on a dielectric core surrounded by the dense nanocrystal quantum dots **c**, schematic presentation of the mechanism for energy transfers between e–h pairs and a resonant plasmon model *via* an exciton level **d**, transmission electron microscope image of a Au core in a Au/silica/dye core–shell nanoparticle structure (**e**), and spaser mode with $\lambda$=525 nm and $Q$=14.8; the inner and outer circles represent the core and the shell layers, respectively. Figure reproduced with permission, **c**, **d** from Ref, ©2008 NPG, **e**, **f** from Ref, ©2009 NPG.

$$D_s = \frac{\lambda}{2\pi}\left[\mathrm{Re}\left(\frac{-\varepsilon_p^2}{\varepsilon_p + \varepsilon_m}\right)^{1/2}\right]^{-1}$$

In the optical region, the size ($D_s$, corresponding to the so-called nanoplasmon) for noble metals, such as silver, gold and copper, is roughly 25 nm. The minimum size of a nanoplasmonic system is determined by the distance electrons move on the surface in a period of the optical wave. This is given by the formula

$$L_{nl} \sim \upsilon_F / \omega$$

where $\upsilon_F$ is the effective Fermi velocity.

The resonance between a surface plasmon and the electromagnetic field also results in the loss of energy due to the dissipative losses (from the imaginary part of the dielectric function). This induces a loss of the plasmonic field with a decay rate $\gamma_p \propto (\mathrm{Im}\varepsilon_p)^{-1}$. For a metal, a quality factor is defined by

$$Q_p = \frac{-\mathrm{Re}[\varepsilon]}{\mathrm{Im}[\varepsilon]}$$

where $\varepsilon$ is the dielectric function of the metal. This provides a criterion that can be used to estimate whether a given substance is a good plasmonic material in an optical region of interest. The quality factors of gold and silver are shown in Figure 8a. In addition, optical radiation also results in the loss of plasmonic field according to the formula,

$$\gamma_p^r = 4\varepsilon_m^{3/2}\left(\frac{\omega_{sp} a}{c}\right)^3 \left[\frac{\partial \mathrm{Re}\varepsilon_p(\omega_{sp})}{\partial \omega_{sp}}\right]^{-1}$$

Surface plasmons can be amplified by the stimulated emission of radiation in the presence of a gain medium, such as dielectric materials containing excited dye molecules. Lasing depends on the presence of two principal conditions—a cavity for the resonant generation of coherent optical modes and medium with gain due to population inversion. A realization, the so-called spaser (surface plasmon amplification by stimulated emission of radiation) consists of small plasmonic metal particles surrounded by a gain medium such that the linewidth of the light emission from the gain medium overlaps with that of plasmon mode. Thus, the losses of energy of the plasmonic particles due to dispersion and radiation can be compensated by light emission of gain medium. The key point is that by using a composite of plasmonic metal particles in a gain producing medium one may obtain a coherent radiation field in a lasing system that is smaller than the wavelength of the light.

Stockman used a quantized treatment of a gain medium and a quasiclassical treatment of the surface plasmon to construct a semiclassical dynamical equation for the spaser. This gives a description of the spaser mechanism as shown in Figure 8c and 8d. The equation of motion of the density matrix of a *p*th chromophore (the active elements in the gain medium) can be expressed as $i\hbar\dot{\rho}^{(p)} = \left[\rho^{(p)}, \hat{H}\right]$, with $\hat{H} = \hat{H}_g + \sum_n \hbar\omega_n \hat{a}_n^+ \hat{a}_n + \sum_p \hat{E}(r_p)\hat{d}^{(p)}$, where $\hat{H}_g$ is the Hamiltonian of gain medium. $r_p$ and $\hat{d}^{(p)}$ are the coordinate vector and dipole moment operators of the *p*th chromophore in gain medium, respectively. The second term in the total Hamiltonian is for the quantized surface plasmon and the third term is for the coupling between the chromophore and the electric field of surface plasmon. By defining the gain factor of chromophore, which is a two energy level system with the formula

$$g = \frac{4\pi\omega\sqrt{\varepsilon_m}}{3c\hbar\Gamma_{12}}\left|d_{12}\right|^2 n_c$$

where $d_{12}$ is the transitional dipole element between ground state and excited state, the threshold gain for spasing can be obtained and expressed as

$$g_{th} = \frac{\omega}{c\sqrt{\varepsilon_m}} \frac{\mathrm{Re}[s(\omega)]}{1 - \mathrm{Re}[s(\omega)]} \mathrm{Im}[\varepsilon_p(\omega)]$$

Here $\Gamma_{12}$ is the rate constant for describing the polarization relaxation, $n_c$ is the density of the chromophores, and $s(\omega) = \varepsilon_m / (\varepsilon_m - \varepsilon_p)$ is Bergman›s spectral parameter. One may note that the threshold for gain just depends on the spasing frequency and the dielectric properties of system. However, the spasing frequency is determined by the geometry of the system.

Progress in spasers, including both theory and experiment, has been rapid. The original theoretical concept of the spaser was proposed with V-shaped metallic structures and semiconductor quantum dots. Following this, a nanolens spaser was proposed with a linear chain structure of metal nanospheres combined with an active medium. A proposal for a spaser based on metal cores with an active shell was considered on basis of linear electrodynamics. A narrow-diversion coherent radiation on based on the combination of a metamaterial and a spaser was proposed by Zheludev *et al*. The combination of the plasmon of an anisotropic spherical particle and an active medium was also proposed to result in a spaser. In experiments, a spaser was realized on a conjugate structure based on a metallic core and a dye-doped dielectric shell (Figure 8). Spasers have also been realized in other nanostructures, such as in CdS nanowires combined with a silver substrate and separated by a $MgF_2$ layer. The development of active subwavelength optical elements such as in spasers, is expected to lead not only to diverse applications, but also to new fundamental insights into nonlinear light matter interactions.

## REFERENCES

1. Andrew P, Barnes WL . Energy transfer across a metal film mediated by surface plasmon polaritons. *Science* 2004; **306**: 1002–1005.
2. Anker JN, Hall WP, Lyandres O, Shah NC, Zhao J *et al.* Biosensing with plasmonic nanosensors. *Nat Mater* 2008; **7**: 442–453.
3. Barnes WL, Dereux A, Ebbesen TW . Surface plasmon subwavelength optics. *Nature* 2003; **424**: 824–830.
4. Bergman DJ, Stockman MI . Surface plasmon amplification by stimulated emission of radiation: quantum generation of coherent surface plasmons in nanosystems. *Phys Rev Lett* 2003; **90**: 027402.
5. Bohren CF, Huffman DR . *Absorption and Scattering of Light by Small Particles*. New York: John Wiley & Sons, Inc.; 1998.
6. Dieringer JA, McFarland AD, Shah NC, Stuart DA, Whitney AV *et al.* Introductory lecture surface enhanced Raman spectroscopy: new materials, concepts, characterization tools, and applications. *Farad Discuss* 2006; **132**: 9–26.
7. Fu Y, Zhang J, Lakowicz JR . Plasmon-enhanced fluorescence from single fluorophores end-linked to gold nanorods. *J Am Chem Soc* 2010; **132**: 5540–5541.
8. Green MA, Pillai S . Harnessing plasmonics for solar cells. *Nat Photonics* 2012; **6**: 130–132.
9. Hutter E, Fendler JH . Exploitation of localized surface plasmon resonance. *Adv Mater* 2004; **16**: 1685–1706.
10. Martin RM . *Electronic Structure: Basic Theory and Practical Methods*. Cambridge: Cambridge University Press; 2004.
11. Nie S, Emory SR . Probing single molecules and single nanoparticles by surface-enhanced Raman scattering. *Science* 1997; **275**: 1102–1106.
12. Rayleigh L . On the scattering of light by small particles. *Philos Mag* 1871; **41**: 447–454.

13. Shannon RD, Shannon RC, Medenbach O, Fischer RX . Refractive index and dispersion of flourides and oxides. *J Phys Chem Ref Data* 2002; **31**: 931–970.
14. Shinji H, Takayuki O . Plasmonics: visit the past to know the future. *J Phys D Appl Phys* 2012; **45**: 433001.
15. Singh DJ . Structure and optical properties of high light output halide scintillators. *Phys Rev B* 2010; **82**: 155145.
16. Willets KA, Van Duyne RP . Localized surface plasmon resonance spectroscopy and sensing. *Ann Rev Phys Chem* 2007; **58**: 267–297.
17. Zayats AV, Smolyaninov II, Maradudin AA . Nano-optics of surface plasmon polaritons. *Phys Rep* 2005; **408**: 131–314.
18. Zhao J, Zhang X, Yonzon CR, Haes AJ, Van Duyne RP . Localized surface plasmon resonance biosensors. *Nanomedicine* 2006; **1**: 219–228.

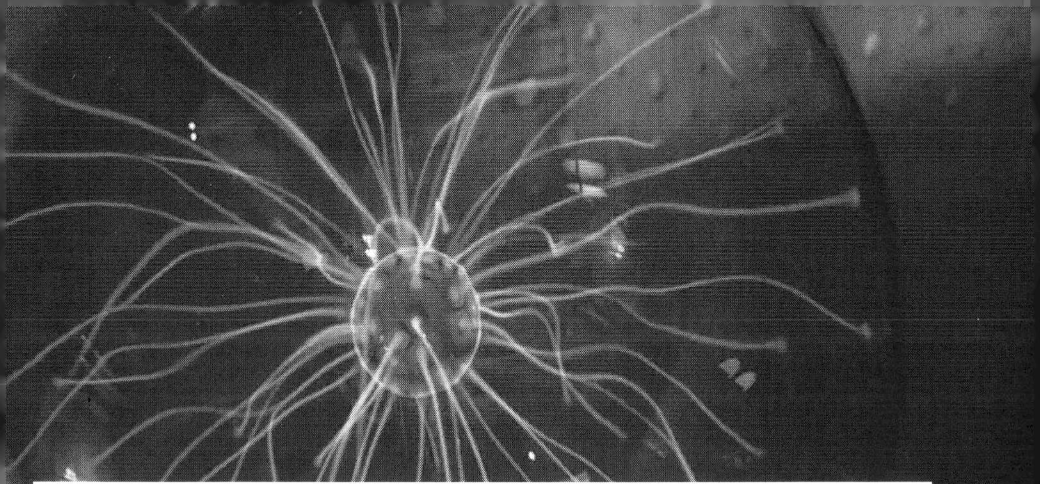

CHAPTER 3

# ELECTROMAGNETIC FIELD ENHANCEMENT IN SURFACE-ENHANCED RAMAN SCATTERING

## INTRODUCTION

Surface-enhanced Raman scattering (SERS) was first observed in1974 by Fleischmann et al. for pyridine, $C_5NH_5$, adsorbed on a silver electrode roughened by means of successive oxidation-reduction cycles. Combining SERS experiments and theoretically obtained electromagnetic (EM) field distribution around metal surfaces, the EM effect has been recognized as the dominating factor contributing to SERS in most cases, compared to the additional about $10^2$ chemical enhancement. In the year 1997, Nie and Emory and Kneipp et al. reported SERS enhancement as large as $10^{14}$ in the experiments for single-molecule SERS. The numerical simulations confirm that the EM enhancement in the nanogaps of nanoparticle dimers can be extraordinarily high and dominate among the factors contributing to single-molecule SERS. The high EM enhancement in the nanogap between the nanowire and the nanoparticle renders single-molecule SERS to be achieved remotely from the laser illumination point.

## 3.1 SURFACE-ENHANCED RAMAN SCATTERING (SERS)

Surface-enhanced Raman scattering (SERS) is a surface-sensitive technique that enhances Raman scattering by molecules adsorbed on rough metal surfaces or by nanostructures such as plasmonic-magnetic silica nanotubes. The enhancement factor can be as much as $10^{10}$ to $10^{11}$, which means the technique may detect single molecules.

Surface-enhanced Raman scattering (SERS) was first observed in 1974 by Fleischmann et al. for pyridine, $C_5NH_5$, adsorbed on a silver electrode roughened by means of successive oxidation reduction cycles. This observation of increased Raman intensity was explained by an increase in surface area, that is, an increase in the number of adsorbed molecules contributing to the Raman signal. But, in 1977, Jeanmaire and Van Duyne and, independently, Albrecht and Creighton repeated these experiments and concluded that the enormously strong Raman signal must be caused by a true enhancement of the Raman scattering efficiency itself. Since then the SERS effect has been observed for many molecules adsorbed on a number of metals. On specially prepared rough metal surfaces or colloidal metal nanoparticles, the Raman signal can be a million-fold more intense than that of free molecules. More recently, strikingly large enhancement factors, over $10^{10}$ in magnitude, have been observed for molecules adsorbed on colloidal silver particles. Such a high enhancement factor renders, for the first time, the Raman spectra of a single molecule to be observed.

Combining SERS experiments and theoretically obtained electromagnetic (EM) field distribution around metal surfaces, the EM effect has been recognized as the dominating factor contributing to SERS in most cases, compared to the additional about $10^2$

chemical enhancement. When light is incident on the surface of a roughened metal substrate, a local EM field with strength much stronger than the incident light is generated in the vicinity of the substrate. The highly magnified local field significantly strengthens the Raman scattering process of molecules adsorbed on the substrate, which works effectively as an antenna capable of collecting photons. Furthermore, the substrate also enhances the Raman scattering light emitted from the molecules, playing a role of an antenna for broadcasting. Thus the EM effect contributes a total enhancement factor of

$$M_{em} \approx [|\vec{E}_L(\omega_I)|^2/|\vec{E}_I(\omega_I)|^2]\cdot[|\vec{E}_L(\omega_I \pm \omega_V)|^2/|\vec{E}_I(\omega_I \pm \omega_V)|^2], \tag{1}$$

where $\omega_I$ is the frequency of incident light, $\omega_V$ is the vibrational frequency of the molecule undergoing the Raman process, $\vec{E}_I$ is the electric field of the incident light, and $\vec{E}_L$ is the local magnified field feeding the molecule. The first term in the square brackets on the right-hand side of Eq. 1 reflects the capability of the substrate to collect incident light, while the second term reflects its capability to enhance the Raman emission from the molecule, where the plus and minus signs in the parentheses correspond to anti-Stokes and Stokes scattering, respectively. Since usually $\omega_V << \omega_I$, the EM enhancement factor can be roughly estimated as (abbreviated as $E^4$)

$$M_{em} \approx |\vec{E}_L(\omega_I)|^4/|\vec{E}_I(\omega_I)|^4. \tag{2}$$

The EM effect originates from surface plasmon (SP) excitation, which is the collective electronic excitation mode manifested as coherent electron density oscillation localized at the surface of conductive materials. The surface roughness enables the SP resonance to focus the light at the subwavelength scale, forming hot spots with strong enhancement of the EM field. On the other hand, when the incident light is used to excite the SPs, energies of the photon and plasmon must match, as well as their momenta. These matching conditions may be simultaneously satisfied when the translational symmetry is broken, which renders surface roughness conducive to coupling light to SP resonances.

Extensive studies of EM enhancement have been focused on metal nanoparticles and their aggregates. The localized surface plasmons (LSPs) excited on nanoparticles usually exhibit a large enhancement factor, with the $E^4$ reaching $10^4$ or higher. Using the Drude model and assuming the applicability of quasi-static approximation, we can make the simple rationale: for a nanosphere, its dynamic electric polarizability is $\alpha(\omega) = R^3\left[1-\frac{\omega}{\Omega}\left(\frac{\omega}{\Omega}+i\frac{\gamma}{\Omega}\right)\right]^{-1}$, where R is the radius of the sphere, $\omega$ and $\Omega$ are the frequencies of incident light and LSPs, respectively, and $\gamma$ measures the effective damping. For incident light with an electric field $\bar{E}_{I}(\omega)$, the local field around the nanosphere is $\bar{E}_{L}(\omega) \approx \bar{E}_{I}(\omega)\cdot\left[1-\frac{\omega}{\Omega}\left(\frac{\omega}{\Omega}+i\frac{\gamma}{\Omega}\right)\right]^{-1}$. Since typically the damping is $\gamma \approx 0.1\ \Omega$ or less, $E^4$ is ~$10^4$ or higher at resonance ($\omega = \Omega$).

The most distinct feature of metal nanoparticles is that when they aggregate, extraordinarily high EM enhancement over $10^{10}$ can be reached in the nanogaps between two particles, rendering single-molecule SERS possible. The extraordinary EM enhancement in nanogaps originates from strong EM coupling between individual nanoparticles, and the coupling strength and enhancement factor rapidly increase when the size of the nanogap decreases. However, when the gap separation goes down to a very small value, smaller than 1 nm, electronic tunneling between the nanoparticles becomes significant, which reduces the interparticle coupling and tends to lower EM enhancement.

### 3.1.1 Mechanisms of the Enhancement Effect of SERS

The exact mechanism of the enhancement effect of SERS is still a matter of debate in the literature. There are two primary theories and while their mechanisms differ substantially, distinguishing them experimentally has not been straightforward. The electromagnetic theory proposes the excitation of localized surface plasmons, while the chemical theory proposes the formation of charge-transfer complexes. The chemical theory is based on resonance Raman spectroscopy, in which the frequency coincidence (or resonance) of the incident photon energy and electron transition greatly enhances Raman scattering intensity. Research in 2015 on a more powerful extension of the SERS technique called SLIPSERS (Slippery Liquid-Infused Porous SERS) has further supported the EM theory.

## *Electromagnetic Theory*

The increase in intensity of the Raman signal for adsorbates on particular surfaces occurs because of an enhancement in the electric field provided by the surface. When the incident light in the experiment strikes the surface, localized surface plasmons are excited. The field enhancement is greatest when the plasmon frequency, $\omega_p$, is in resonance with the radiation ($\omega = \omega_p / \sqrt{3}$ for spherical particles). In order for scattering to occur, the plasmon oscillations must be perpendicular to the surface; if they are in-plane with the surface, no scattering will occur. It is because of this requirement that roughened surfaces or arrangements of nanoparticles are typically employed in SERS experiments as these surfaces provide an area on which these localized collective oscillations can occur. SERS enhancement can occur even when an excited molecule is relatively far apart from the surface which hosts metallic nanoparticles enabling surface plasmon phenomena.

The light incident on the surface can excite a variety of phenomena in the surface, yet the complexity of this situation can be minimized by surfaces with features much smaller than the wavelength of the light, as only the dipolar contribution will be recognized by the system. The dipolar term contributes to the plasmon oscillations, which leads to the enhancement. The SERS effect is so pronounced because the field enhancement occurs twice. First, the field enhancement magnifies the intensity of incident light, which will excite the Raman modes of the molecule being studied, therefore increasing the signal of the Raman scattering. The Raman signal is then further magnified by the surface due to the same mechanism that excited the incident light, resulting in a greater increase in the total output. At each stage the electric field is enhanced as $E^2$, for a total enhancement of $E^4$.

The enhancement is not equal for all frequencies. For those frequencies for which the Raman signal is only slightly shifted from the incident light, both the incident laser light and the Raman signal can be near resonance with the plasmon frequency, leading to the $E^4$ enhancement. When the frequency shift is large, the incident light and the Raman signal cannot both be on resonance with $\omega_p$, thus the enhancement at both stages cannot be maximal.

The choice of surface metal is also dictated by the plasmon resonance frequency. Visible and near-infrared radiation (NIR) are used to excite Raman modes. Silver and gold are typical metals for SERS experiments because their plasmon resonance frequencies fall within these wavelength ranges, providing maximal enhancement for visible and NIR light. Copper's absorption spectrum also falls within the range acceptable for SERS experiments. Platinum and palladium nanostructures also display plasmon resonance within visible and NIR frequencies.

## *Chemical Theory*

Resonance Raman spectroscopy explains the huge enhancement of Raman scattering intensity. Intermolecular and intramolecular charge transfers significantly enhance Raman spectrum peaks. In particular, the enhancement is huge for species adsorbing the metal surface due to the high-intensity charge transfers from the metal surface with wide band to the adsorbing species. This resonance Raman enhancement is dominant in SERS for species on small nanoclusters with considerable band gaps, because surface plasmon appears only in metal surface with near-zero band gaps. This chemical mechanism probably occurs in concert with the electromagnetic mechanism for metal surface.

### 3.1.2 Applications of SERS

SERS is highly sensitive analytical spectroscopic technique available among the modern scientific communities and can contribute both to surface science and nanoscience. It is also associated with a broad range of other surface sensitive techniques to study various fundamental and applied research topics such as, corrosion, catalysis, advanced materials, diagnostics, biomedical applications, biological process, and sensing.

The SERS-active Ag/AAO nanostructured system has been used by Liu et al. to investigate antibiotic-induced chemical changes in bacterial cell wall. They recorded high quality, highly reproducible SERS spectra, which are also sensitive and stable. The "chemical features" observed from the SERS spectra of bacterial cell wall facilitates fast identification of drug-resistant bacteria within an hour both qualitatively and quantitatively. Furthermore, the characteristic changes in the SERS spectra were clearly observed in the drug-sensitive bacteria at the near the beginning (i.e., 1 hour) of antibiotic exposure, which could be utilized to differentiate them from the drug resistant ones. The rapid detection of pathogens such as bacteria and viruses using the SERS technique provides a novel approach for microbial diagnostics. The SERS-based novel technique was applied to a single bacterium. This rapid SERS detection of pathogens makes possible direct analysis of clinical specimen as an alternative to a pure culture specimen. Traditional diagnostic protocols for diagnosing bacterial infections are based on the isolation of pure culture of the bacterium, which is followed by the resolving of the nature of the isolate and an extensive assessment of the isolates responses to several antibiotics proliferation or viability. For such biological assays, an incubation period varying from days to weeks or even months is essential for the growth of bacteria with such a density that can be handled by the available diagnostic tools. Different PCR-based protocols have been inducted for the quantitative identification of bacteria. Mass spectrometry can also be alternatively utilized for culture-free bacterial diagnostics. Nevertheless, just like the PCR-based method, mass spectrometry-based protocol is dependent on the existing previous information in the literature about the pathogens. Finally, neither of the PCR or mass spectrometry-based protocols can be applied to a live bacterial sample to observe their responses to antibiotics or to carry out different functional tests. On the other hand, SERS-based spectroscopic technique can resolve the limitation of PCR-based identification methods for pathogens. The SERS active substrates based on the Ag/AAO system can be utilized for the fine changes observed in the bacterial cell wall during different stages of bacterium's growth and also for the bacterium's response to antibiotic treatment during the early period of antibiotic exposure.

In a recent development, Ankamwar et al. fabricated a highly stable and almost homogeneous SERS active substrate from silver nanoparticles synthesized from the leaf extract of *Neolamarckia cadamba* for the rapid detection of two strains of bacteria, Gram-positive (*Staphylococcus aureus*) and Gram-negative (*Escherichia coli*) bacteria. Figure below demonstrates the TEM image of the as-synthesized silver nanoparticles along with their UV-visible spectrum, SAED pattern, and the resultant SERS spectra generated upon interaction with *S. aureus* and *E. coli*.

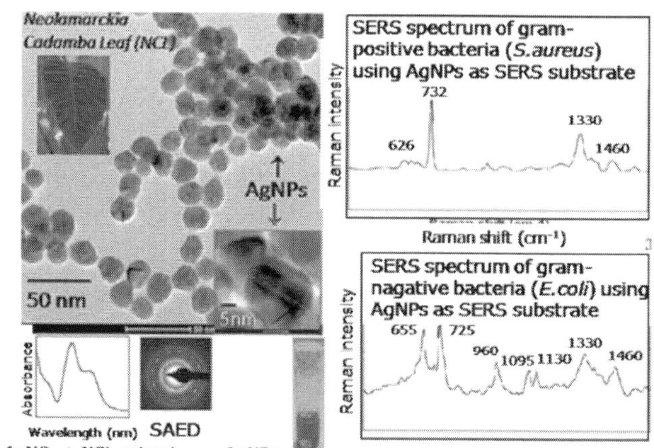


These silver nanoparticles upon interaction with bacteria can exhibit a large Raman enhancement factor (($(3 \pm 0.20) \times 10^7$ and $(5 \pm 0.40) \times 10^7$ for *S. aureus* and *E. coli* bacteria, respectively) with almost zero fluctuations. The SERS substrate developed by them is almost homogeneous with a relative standard deviation value of 6.32 calculated from 50 repeated measurements from various locations on the SERS substrate. In addition to this, the fabricated SERS substrates are extremely stable even after 3 months. Using this almost homogeneous, a stable SERS active substrate, Gram-positive bacteria can be differentiated from Gram-negative bacteria. The SERS data presented in the above study is highly stable, uniform, and reproducible, which shows the versatility of the biosynthesized SERS active substrate. This SERS active substrate is capable of detecting extremely low concentrations ($10^3$ CFU $ml^{-1}$) of *E. coli* within a very short time of 1–5 s and also exhibits high sensitivity. Figure below demonstrates the SERS calibration curve obtained with SERS intensity of the peak at 1330 $cm^{-1}$ (C–N stretching mode) as a function of concentration of bacteria *E. coli*. The 1330 $cm^{-1}$ peak became detectable at $10^3$ CFU/ml of *E. coli* concentration. The SERS intensity increases with the concentration of the bacterial solution, as it is exponentially correlated to the concentration of *E. coli* bacterial cells in the sample between $10^3$ and $10^8$ CFU/ml. Experiments were repeated five times with each bacterial concentration, and the standard errors of the mean for each concentration are also shown in figure below. The major intention of this SERS study using biosynthesized Ag nanoparticles was to develop a rapid fingerprinting method for the characterization of bacteria particularly *E. coli*, which is associated with urinary tract infection (UTI), a common disease among most people of all age groups in developed countries such as India and China.

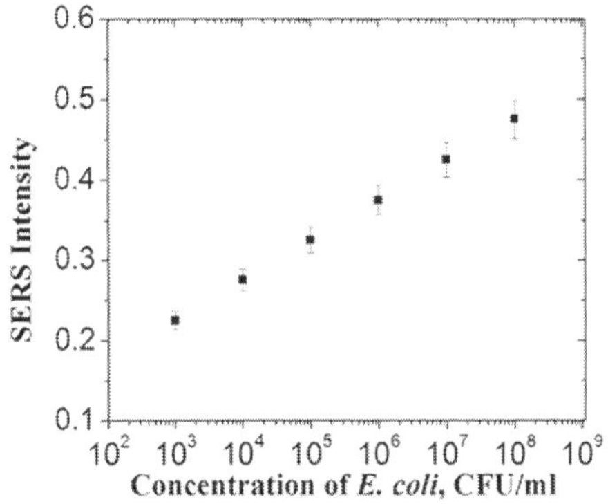

The SERS-based pathogen detection method is especially useful for the analysis of slow-growing bacteria, which typically may take few weeks during laboratory tests. The SERS spectra described above to detect and quantify bacteria lacks the molecular level specificity compared to other commonly used techniques. The SERS-based technique demonstrates a novel approach for rapid microbial diagnostics, where SERS can be directly applied on the clinical sample rather than pure cultured bacteria.

Nie and Emory carried out a single-molecule SERS experiment by employing the SERS technique along with the transmission electron microscopy (TEM) and scanning tunneling microscopy (STM) techniques. They observed surface Raman enhancement in the order of $10^{14}$–$10^{15}$ for single rhodamine 6G (R6G) molecule adsorbed on selected Ag nanoparticles. For the single-molecule SERS study, a single event was observed rather than an ensemble averaged value usually attained for traditional SERS measurements. Advancement of the single molecule SERS technique has brought a new aspect in biomedical research, as it can act as a versatile probing tool to investigate various biological molecules such as virus, bacteria, and protein.

The limitation arises out of surface generality of the SERS effect has been resolved by the invention of tip-enhanced Raman spectroscopy (TERS) technique in 2004, which is a modification of the conventional SERS technique. This new and novel technique is derived from the enhancement of the surface Raman scattering intensity (SERS signal) by merging Raman spectroscopy with a scanning probe microscopy technique such as atomic force microscopy (AFM) or scanning tunneling microscopy (STM). The experimental setup in the TERS technique consists of AFM or STM tip placed in a nearby surrounding area of an ultrasmooth substrate, generally single crystal metal surfaces and illuminated by an electromagnetic radiation of suitable wavelength. The contact or tunneling mode of AFM/STM can be used in the experimental setup of TERS. Irradiating with a laser beam of suitable wavelength, a localized surface plasmons are excited in the tip-substrate gap, generating a huge, local enhancement of electromagnetic field in comparison to the incident radiation. TERS was used to probe malachite green isothiocyanate, a dye molecule adsorbed on the Au(111) surface. TERS has been used to study surface reactions on single crystal and smooth surfaces, as surface roughness of the substrate does not play any role in this enhancement.

Surface-enhanced Raman scattering spectroscopy (SERS) can be used for the identification short-live reaction intermediates such as radical and radical ions on the electrode surface and elucidation of the reaction mechanism in general. Tian and his research group at Xiamen University, China, carried out the first *in situ* electrochemical SERS (EC-SERS) investigation on the electrochemical reduction of $PhCH_2Cl$ in acetonitrile ($CH_3CN$) on the Ag electrode. The benzyl radical anion as a reaction intermediate and 3-phenylpropanenitrile as the major reaction product were detected from the SERS study for the above surface reaction. The complete reaction mechanism enlightening the adsorption process of $PhCH_2Cl$ on the Ag surface and all other possible interactions including the solvent molecule have been determined from the systematic SERS study. The SERS results were further validated by quantum mechanical density functional theory (DFT) calculations, which confirm the detection of the reaction intermediate and products.

It was established by Mulvihill et al. that LB assemblies made of various polyhedral Ag nanocrystals can be utilized as high quality SERS active substrates for the high-sensitivity detection of arsenate and arsenite ions in aqueous solutions with a detection limit of 1 ppb. The detection limit resulted from the analysis carried out by the SERS-based technique is an order of magnitude lower than the existing yardstick set by the World Health Organization (WHO). The SERS substrate can be used as a chemical sensor, which is simultaneously highly reproducible and portable, and therefore, this could be easily executed in field detection. The SERS technique can be further employed in environmental analysis. Pesticides, herbicides, pharmaceutical chemicals in water, banned food dyes, aromatic chemicals in regular aqueous solutions and in sea water, chlorophenol derivatives and amino acids, chemical warfare species, explosives, and various organic pollutants can qualitatively and quantitatively analyzed by the SERS-based detection technique. The partition property of SERS substrates as well as surface chemistry facilitates the complete separation of pollutants and analysis of complex environmental samples in real environmental analysis and monitoring.

Immobilized metal nanoparticles in the form of SERS substrates can be used for biomedical diagnostics. For instance, the SERS substrate can be used as a glucose sensor to detect glucose in human blood. Although glucose is most commonly monitored by electrochemical-based sensors, a substitute protocol using SERS substrates fabricated by the NSL technique has been employed to detect glucose in blood. In this new protocol, the SERS-based glucose sensor was developed by growing silver film over nanospheres (AgFON) surfaces prepared by the NSL technique. Nevertheless, glucose sensing on a bare AgFON surface was not successful and glucose was brought within the range of electromagnetic enhancement of the AgFON surface by formation of a self-assembled monolayer (SAM) on its surface to partition the analyte of interest, in a manner similar to the technique used to generate the stationary phase in high-performance liquid chromatography. Numerous SAMs were studied to partition glucose effectively to the AgFON surface and it was observed that both straight-chain alkanethiols and ethyleneglycol-terminated alkanethiols partitioned glucose most effectively. The key for the detection of glucose by the SERS-based technique was the surface chemistry of alkanethiol molecule on the AgFON surface. The SERS substrate was modified with an alkanethiol partition layer to facilitate the glucose adsorption to the metal surface. Real-time sensing and quantitative detection of glucose in bovine plasma by SERS has been reported in the literature.

## 3.2 NUMERICAL APPROACHES TO EM ENHANCEMENT

Because hot spots extend the typical size of only a few nanometers, the EM field within their range cannot be easily probed. Various theoretical and computational approaches capable of evaluating the spatial distribution of EM enhancement around nanostructures with high accuracy have been developed in the last few decades. For nanostructures with all three dimensions larger than a few nanometers, nonlocal effects are negligible and local macroscopic permittivity $\varepsilon(\vec{r},\omega)$ and permeability $\mu(\vec{r},\omega)$ are sufficient for the EM modeling. For these geometries, classical approaches such as finite-difference time-domain (FDTD), generalized Mie theory (GMT), and discrete dipole approximation (DDA) are the most popular ones to simulate the local field distribution. If one or more of the dimensions are small, nonlocal response should be considered and in this regard, first-principle linear response theory is an excellent solution. More recently, McMahon et al. incorporated nonlocal response into the fully retarded classical EM simulation by adopting the hydrodynamic model for permittivity, which effectively improved the simulation of local EM enhancement around apexes of nanostructures.

The FDTD method discretizes and evolves Maxwell's equations both in space and in time. After propagating the EM wave for a sufficiently long time, Fourier transform is performed to obtain the field distribution in the frequency domain. The discretization and evolution of Maxwell's equations implement the Yee algorithm, for which all the components of the electric field $\vec{E}$ and magnetic field $\vec{H}$ are spatially offset from one another, and time-wise $\vec{E}$ and $\vec{H}$ are mutually offset by a 1/2 time step. Different boundary conditions can be imposed on the simulation region, including the absorbing boundary condition used to model wave-scattering problems and the periodic boundary condition applied to systems containing translational symmetry. FDTD is a universal method applicable to various types of optical response problems, including LSPs of nanoparticles and propagating SPs along nanowires.

GMT is based on Mie theory, which analytically describes the scattering of an EM wave by a single spherical object, and extends to calculate the EM response from aggregates of spheres. Similar to Mie theory for a single sphere, in GMT both incident and scattered fields are expanded in terms of vector spherical harmonics (VSH) $|mnsp\rangle$ serving as the basis. Here m and n are angular momentum quantum numbers; s = 1 for the incident wave and 3 for the scattered wave, which satisfies different boundary conditions in faraway space; and p = 1 for the transverse electric field $\vec{M}$ and 2 for the transverse magnetic field $\vec{N}$. The great advantage of GMT is that its accuracy does not change with the size of a sphere and it is a highly precise approach to simulate aggregates of spherical nanoparticles. To take account of the interaction between nanospheres, the coordinate transformation for VSH centered at different sites has been developed, which renders a rigorous mathematical connection between incident and scattered waves for different spheres.

DDA divides the nanoparticles into small cubes, each of which has its own electric polarizability determined by the local permittivity (usually the material concerned for SERS is nonmagnetic). The small cubes couple to each other through dipolar interaction, and the electric field radiated from the dipole $\vec{P}_j$ and reaching position $\vec{r}_{ij}$ is

$$\vec{E}_{ij} = k^2 e^{ikr_{ij}} \frac{\vec{r}_{ij} \times (\vec{r}_{ij} \times \vec{P}_j)}{r_{ij}^3} + e^{ikr_{ij}} (1 - ikr_{ij}) \frac{[r_{ij}^2 \vec{P}_j - 3\vec{r}_{ij}(\vec{r}_{ij} \cdot \vec{P}_j)]}{r_{ij}^5}. \quad (3)$$

Eventually one ends up with a set of self-consistent equations for all the dipoles constituting the nanoparticles in the simulated system.

McMahon et al. recently developed a technique to take account of the spatial nonlocal response of materials by incorporating the wave vector $\vec{k}$-dependent permittivity $\varepsilon(\vec{k},\omega)$ into the computational EM methods. This development significantly improves the calculation of EM response from nanostructures consisting of dimensions smaller than 10 nm. In one example, they showed that when nonlocal effects are considered, the field enhancements around the apexes of gold triangular nanowires are significantly lower than the results obtained from local electrodynamics.

A more accurate method for low dimensions, where quantum effects are significant, is to combine linear response theory with electronic structure calculations for the nanostructures. The noninteracting density–density response function is in the following form:

$$\chi(\vec{r},\vec{r}',\omega)=\sum_{i,j}(f_i-f_j)\frac{\psi_i(\vec{r})\psi_j^*(\vec{r})\psi_j(\vec{r}')\psi_i^*(\vec{r}')}{\omega-\omega_{ij}+i\eta}, \quad (4)$$

where $f_i$ is the Fermi occupation number for energy $\varepsilon_i$, $\omega_{ij}=(\varepsilon_i-\varepsilon_j)/\hbar, \eta\rightarrow 0^+$, and $\Psi_i$ is the single-body electronic wave function. The induced charge density that determines the local EM enhancement is calculated as

$$\rho_{\text{ind}}(\vec{r},\omega)=\int\chi(\vec{r},\vec{r}',\omega)\cdot\Phi_{\text{tot}}(\vec{r}',\omega)d^3r', \quad (5)$$

where $\Phi_{tot}$ is the distribution of total electric potential. Applying the electrostatic relation $\Phi_{tot}=\Phi_{ext}+\Phi_{ind}$ and $4\pi\varrho_{in}d=-\nabla^2\Phi_{ind}$, where $\Phi_{ext}$ and $\Phi_{ind}$ are the electric potentials of the external field and the induced field, respectively, $\Phi_{ind}$ and $\varrho_{ind}$ can both be selfconsistently calculated. The local EM enhancement is calculated as $M=|\nabla\Phi_{tot}|^4/|\nabla\Phi_{ext}|^4$.

## 3.3 THE NANOGAP EFFECT

In the year 1997, SERS enhancement as large as $10^{14}$ in the experiments for single molecule SERS. However, a clear physical picture for the origin of this large enhancement is absent. The huge enhancement originates from the nanogaps of aggregates composed of at least two metal nanoparticles, by positioning single nanoparticle structures to correlate the SERS signals and structural morphologies obtained from scanning electron microscopy (SEM) images. Figure 1 shows the SEM images of silver (Ag) nanoparticles, which aggregate and form dimers, and the insets of Fig. 1c and 1d are Raman spectra of single hemoglobin (Hb) molecules. The single-molecule limit is achieved by decreasing and thus limiting the number of molecules adsorbed to each aggregated nanoparticle. Direct evidence for the spectra being at the singlemolecule level is their temporal fluctuation, which cannot be the result of the ensemble average over a large number of molecules.

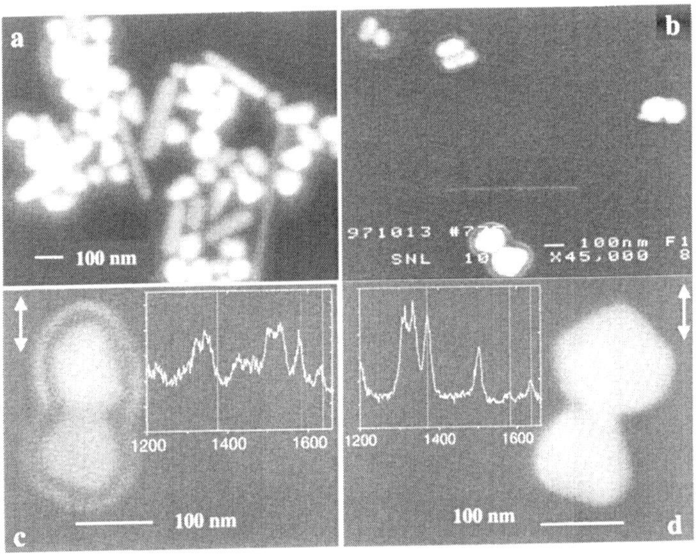


**Figure 1**. SEM images of immobilized Ag nanoparticles. The pictures show (a) an overview of Ag particle shapes and sizes, (b) Ag particle dimers observed after incubation with $1 \times 10^{-11}$ M Hb molecules for 3 h, and (c, d) hot dimers and corresponding single-Hb-molecule spectra. The double-headed arrows in (c) and (d) indicate the polarization of the incident laser field.

The numerical simulations confirm that the EM enhancement in the nanogaps of nanoparticle dimers can be extraordinarily high and dominate among the factors contributing to single-molecule SERS. Figure 2 shows the simulated enhancement spectra for Ag nanosphere dimers and a single nanosphere. As can be seen, the $E^4$ enhancement factor (M) for the dimer is about 4 orders of magnitude larger than for the single nanosphere when the external electric field is polarized along the dimer axis.

The polarization dependence of the enhancement is consistent with the experimental results discussed below. For a nanogap size $d = 5.5$ nm, approximating the dimension of a Hb molecule, the maximal enhancement (M) is about $10^7$ for the wavelengths in experiments. Decreasing the nanogap size from 5.5 to 1 nm results in an increase of M from ~$10^7$ to ~$10^{10}$. The experimentally estimated enhancement factor of ~$10^{10}$ is well within the reach of the EM model without considering the chemical enhancement. It is noted that later studies showing an enhancement factor of $10^7$ can be sufficient for single-molecule SERS detection, and the large enhancement factor of $10^{14}$ claimed in the literature may be an overestimation.

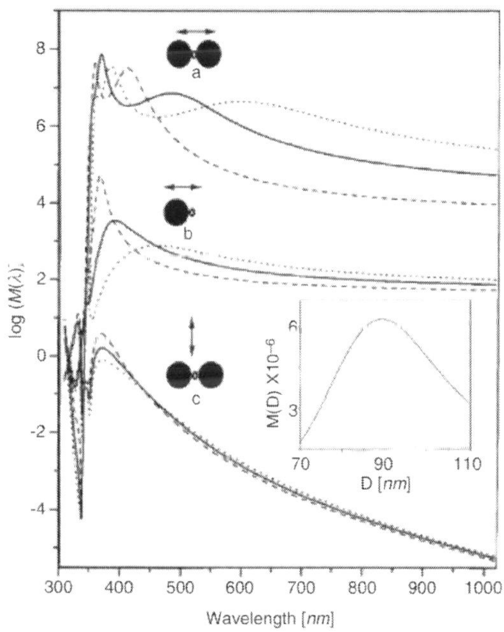


**Figure 2.** Calculated EM enhancement factor for the midpoint between two Ag spheres separated by d = 5.5 nm and for a point d/2 outside a single sphere. The solid and open circles indicate the positions of the Ag spheres and the Hb molecule, respectively. The double-headed arrows indicate the incident polarization. The calculations have been performed for spheres of diameters D = 60 (dashed curves), 90 (solid curves), and 120 nm (dotted curves). The inset shows the enhancement versus D for

an incident wavelength of 514.5 nm and a Stokes shift of 1500 $cm^{-1}$ for configuration a.

Figure 3 shows the calculated spatial distribution of the $E^4$ enhancement factor for single Ag nanoparticles (first row) and dimers with different nanogap sizes (second and third rows). The enhancement in the nanogap of a dimer is orders of magnitude higher than twice of the enhancement near a single nanoparticle because of the strong EM coupling between two nanoparticles. When a pair of closely spaced nanoparticles is excited by incident light, their LSP modes are manifested as electric multipoles and strongly polarize each other, leading to a significantly enhanced induced field compared to that around a single nanoparticle. The mutual polarization is strongest in the nanogap region, which, as a result, contains the highest EM enhancement. The nanoparticles in the right column of Fig. 3 are modeled as rotationally symmetric polygons. As can be seen, the introduction of edges to the nanoparticle does not strongly modify either the magnitude of the enhancement or the spatial distribution, which indicates it is reasonable to use nanospheres to model faceted nanoparticles in experiments.

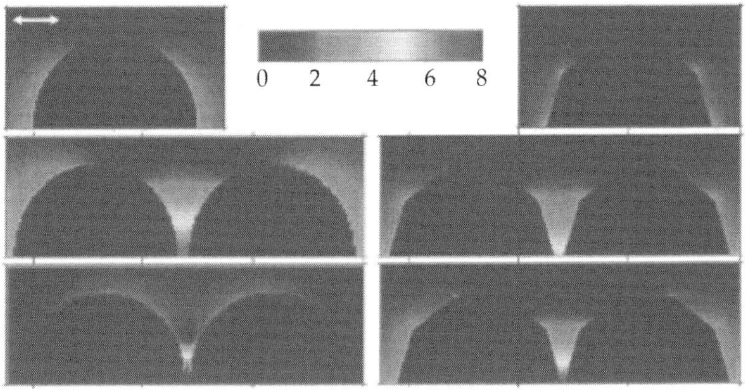


**Figure 3.** EM enhancement factor at a cross section through six different silver particle configurations. The incident field has a wavelength of 514.5 nm with horizontal polarization. The left-hand column shows the EM enhancement for a nanosphere (top) and a dimer of two nanospheres with a separation of 5.5 nm (middle) and 1 nm (bottom). The right-hand column shows the corresponding results for polygons. All

particles share a common largest dimension of 90 nm. Note that the color scale from dark blue to dark red is logarithmic, covering the interval $10^0 < M < 10^8$. Regions with enhancement outside this interval are shown in dark blue and dark red.

The EM enhancement in the nanogap sensitively depends on two factors, the size of the nanogap and the polarization of the external electric field. When the electronic waves of two nanoparticles are well separated, each particle remains neutral during the plasmon excitation. With the decreasing nanogap size, the EM enhancement rapidly increases because of a stronger interparticle coupling (comparing the second and third rows in Fig. 3). Figure 4 shows the simulated EM enhancement spectra in the nanogaps of model nanosphere dimers and single nanospheres with different geometries. The enhancement in the nanogaps of dimers is several orders of magnitude larger than that of single spheres over the whole visible-to-near-infrared spectral range. For both silver and gold nanospheres, when the nanogap size of the dimers increases from 1 to 5.5 nm, the enhancement dramatically drops by several orders of magnitude over the whole visible-to-near-infrared spectral range.

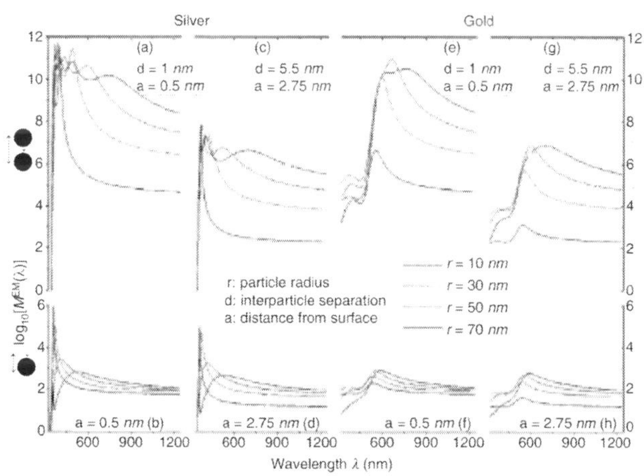


**Figure 4**. The upper row shows the EM enhancement factor $M^{EM}$ versus the wavelength at the midpoint between two equal-radius spherical particles, composed of silver (a, c) and gold (e, g), separated by a gap of d = 1 nm (a, e) and d = 5.5 nm (c, g) for various radii r. When one particle is removed the EM enhancement factor decreases dramatically, as illus-

trated by the data for single particles shown in the lower row (b, d, f, h). Calculations have been performed for the polarization parallel to the dimer axis and for the surrounding medium of vacuum. The dielectric constants of the particles are based on the experimental data of Johnson and Christy.

On the other hand, for a specific dimer, the highest EM enhancement is obtained when the external field is polarized along the dimer axis. Therefore, the intensity of the SERS signal is dependent on the laser polarization and the strongest for polarization along the dimer axis. Actually, the enhancement has approximately a $\cos^2\alpha$ dependence on the angle $\alpha$ between the polarization and the dimer axis. This is because the local EM field enhancement in the $E^4$ law has a $\cos^2\alpha$ polarization dependence, while the Raman emission enhancement is independent of angle $\alpha$. Figure 5 illustrates the experimental results of the polarization dependence of SERS in the dimers, and Fig. 6 shows the simulation results of the polarization dependence of EM enhancement. The consistency between Figs. 5 and 6 again supports that the nanogap effect is the key ingredient of ultrasensitive SERS analysis.

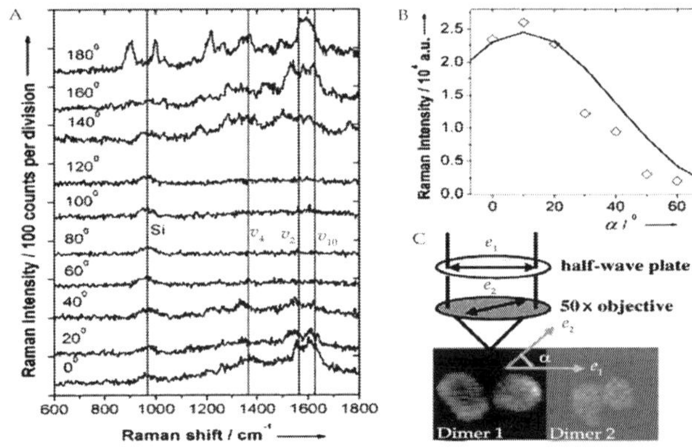


**Figure 5**. Polarization-dependent SERS spectra (A) and SERS intensity (B) from the corresponding dimers shown in the SEM images (C). Dimer 1: 285 × 285 nm$^2$ for (A) and Dimer 2: 420 × 420 nm$^2$ for (B). The half-wave plate in the inset rotates the polarization vector of the excitation beam ($e_1$) by an angle $\alpha$. In (A), the vertical lines mark the heme modes ($v_2$, $v_4$, and $v_{10}$) and the twophonon band of the Si substrate. In (B), the vertical axis shows the integrated intensity between 700 and 2200 cm$^{-1}$ Stokes

shifts; ◊ experimental data, — fit to $\cos^4(\alpha - \alpha_0)$. All measurements were performed using excitation at 514.5 nm and with the same collection time (30 s). The incident irradiance was ~1 μW/μm² in (A) and ~0.1 μW/μm² in (B). The angle $\alpha_0$ between the dimer axis and the $e_1$ axis was ~0° in (A) and 10° in (B).

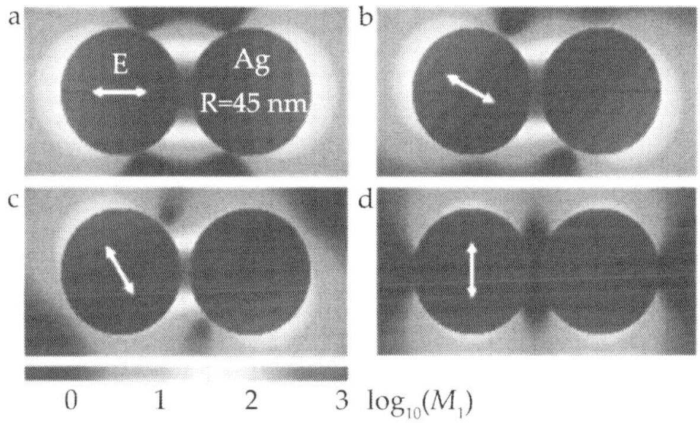


**Figure 6**. Local intensity enhancement $M_1 = (E_{loc}/E_0)^2$ in logarithmic scale in a plane through the centers of the Ag nanospheres and perpendicular to the incident wave vector for different incident polarizations indicated by the arrows: (a) 0°, (b) 30°, (c) 60°, and (d) 90°.

Figures 2 and 4 also show that the peak positions of EM enhancement in the nanogap of a dimer and around a single nanoparticle are significantly different. The difference originates from EM coupling between nanoparticles, which leads to the hybridization of plasmon modes belonging to individual nanoparticles. As a result of the hybridization, new modes are formed and shifted significantly from the original ones. Figure 7 schematically illustrates the plasmon hybridization, where l indexes the angular momentum of the plasmon modes of individual nanoparticles. The plasmon hybridization is in analogy to the combination of atomic orbitals, which forms molecular orbitals with a new set of energy levels.

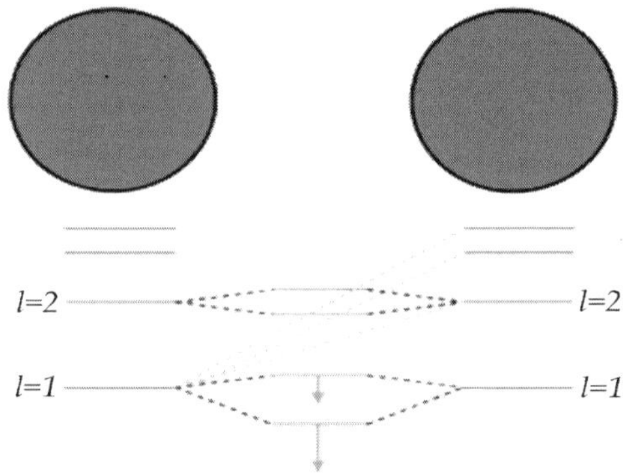


**Figure 7**. Schematic picture illustrating the plasmon hybridization in a nanoparticle dimer. The individual nanosphere plasmons on the two particles interact and form bonding and antibonding dimer plasmons. In the dimer geometry, nanosphere plasmons with a given angular momentum l interact also with plasmons of a different angular momentum on the other particle. This interaction induces extra shifts of the dimer plasmons at small separations. The figure illustrates this shift for the l = 1 derived dimer plasmons.

The hybridization of plasmon modes will lead to the change of plasmon resonance frequencies that can be observed from extinction spectra. Figure 8 shows the calculated EM enhancement spectra and extinction/scattering/absorption spectra for Ag nanoparticle dimers and pentamers. For the dimer, the long-wavelength peak from the coupling of the dipoles of the two particles is prominent in both the EM enhancement spectra of the nanogap site and the extinction spectra for small polarization angles (0° and 30°). The short-wavelength peaks in the extinction spectra for 0° and 30° are mainly from quadrupolar resonances. When the polarization is perpendicular to the dimer axis, the coupled dipolar mode shifts to short wavelength with low strength, and the EM enhancement in the nanogap more or less vanishes. The accordance between the near-field EM enhancement and far-field extinction can be seen from Fig. 8a. For the pentamer, however, the results are quite different, as shown in Fig. 8b. The extinction spectra are dominated by a broad long-wavelength peak of dipolar origin. In contrast, the nearfield intensities in the nanogaps exhibit resonances much sharper than for the dimer and are extremely polarization dependent. The narrower resonances are expected to originate from the coherent superposition of the EM field from different particles. The results for the pentamer indicate that the properties of near field and far field for the same aggregate can be distinctly different. The numerous nanogaps in the multiple-particle aggregate may become "hot" for different wavelengths and different polarizations. It is noted that the resonance peaks in the spectra show asymmetric profiles, which originate from the interference of dipolar and high-order modes, that is, the so-called Fano resonances.

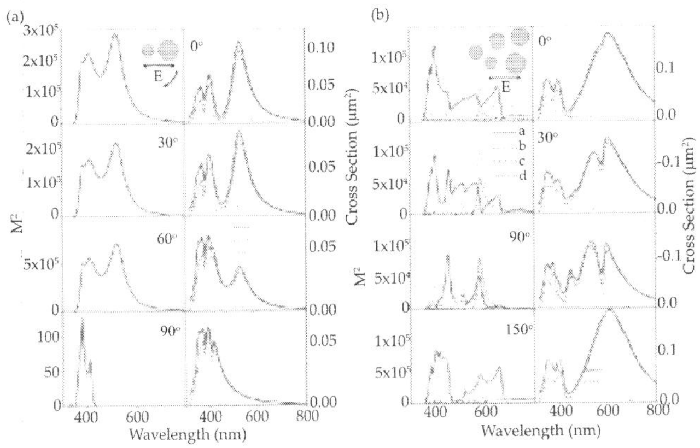


**Figure 8**. Near-field intensity enhancement $M^2(\lambda)$ at hot gap sites and farfield cross sections (extinction, scattering, absorption) for a dimer (diameter D – 60 nm and 100 nm) and a pentamer (D – 60 nm, 70 nm, 80 nm, 90 nm, and 100 nm) in vacuum with gap dimensions d = 1 nm throughout. The spheres are situated on a planar virtual surface, and the incident wave vector is normal to this surface, with polarization as indicated in the insets. The calculations included multipoles up to order N = 30 and scattering processes up to infinite order through the matrix inversion technique.

## 3.4 VARIOUS TYPES OF NANOGAPS

In addition to aggregates of nanoparticles, nanogaps favored by Raman scattering are also formed in several other types of systems. One example is tip-enhanced Raman scattering (TERS), with the molecules trapped in the nanogap between a metal tip and a metal substrate. The configuration is schematically illustrated in Fig. 9. According to the numerical simulation carried out by Yang et al., when the distance between the gold tip and the gold substrate is d = 2 nm, their strong near-field EM coupling results in $1.8 \times 10^9$ EM enhancement for TERS in the nanogap. The coupling strength in TERS is also strongly dependent on the tip–substrate separation since the EM near field decays in an exponential manner far from the surface. On the other hand, a sharp tip with a small radius produces a relatively large enhancement and high spatial resolution of TERS, which is due to the lightning rod effect resulting from the increasing confinement of the surface charges at the sharp tip apex. Figure 10 shows the simulated maximum EM enhancement in the nanogap for different tip radii and the spatial distribution of enhancement for a gold tip with a radius of 25 nm. When the radius decreases from 10 to 5 nm, the enhancement in the nanogap is rapidly increased. The spatial resolution quantified as the full-width at half-maximum (FWHM) of the Raman enhancement is less than 10 nm for the 25 nm radius gold tip.

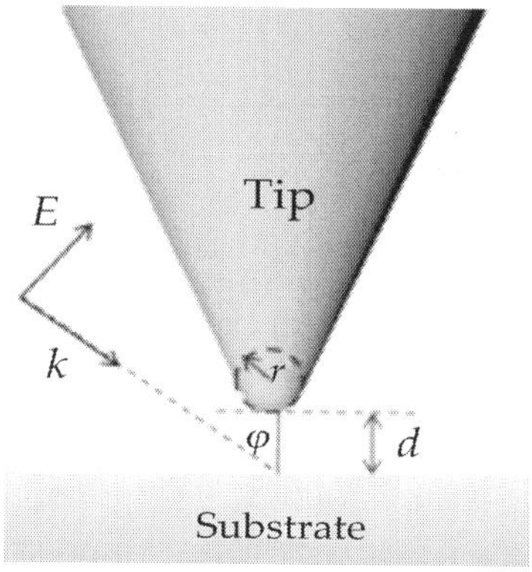


**Figure 9**. Schematic diagram of the metal tip and substrate geometry: the tip is modeled as a conical taper terminated by a hemisphere of radius r, held at a distance d from the substrate surface. An EM plane wave is incident at an angle $\phi$, with the polarization in the plane of incidence.

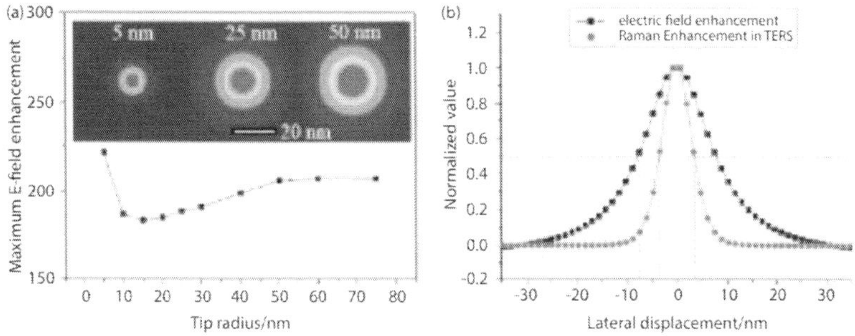


**Figure 10**. Tip size dependence of the maximum field enhancement and spatial resolution for a gold tip held at 2 nm above a gold substrate surface. The 632.8 nm plane wave is incident from the side at an angle of 60° with respect to the substrate normal, with the polarization in the plane of incidence. (a) Maximum electric field enhancement in the nanogap as a function of the tip radius. The insets show the spatial field distribution in a plane parallel to and 1 nm above the substrate for tip radius r = 5, 25, and 50 nm, from left to right. (b) Normalized electric field enhancement M and Raman enhancement $M^4$ along a horizontal line 1 nm below the apex of a gold tip with a 25 nm radius above a gold substrate.

Another type of nanogap is formed between a metal nanoparticle and a nanowire, and the particle–wire coupling results in a strong EM enhancement, showing a SERS signal comparable to those in the nanogaps of nanoparticle dimers. Figure 11a shows the experimental SERS spectra of malachite green isothiocyanate (MGITC) molecules at different spots of the coupled system. Clearly the highest enhancement is in the junction between the nanowire and the nanoparticle. The EM simulation shows that for the 5 nm gap separation between the particle and wire, the average SERS enhancement factor is of the order of $10^6$ for the different shapes of particles and the maximum local SERS enhancement for perpendicular polarization is of the order of $10^{10}$, which is similar to the maximum enhancement in the nanogap of nanoparticle dimers. The polarization dependence of the EM enhancement in the nanogap can be clearly seen from Fig. 11b.

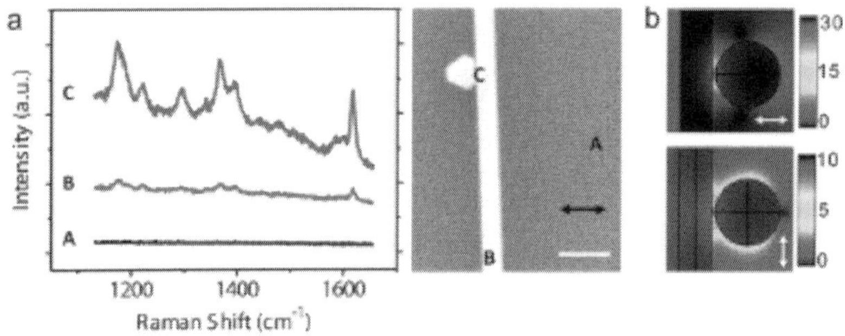


**Figure 11**. (a) Raman spectra of MGITC from different positions of the sample. The arrow in the SEM image shows the incident polarization. The scale bar is 400 nm. (b) Calculated electric field for a gold sphere of radius 50 nm at 5 nm from a wire of radius 25 nm for polarization perpendicular (top) and parallel (bottom) to the wire.

For an infinitely long metal nanowire, which is the ideal model for a long enough wire in reality, the plasmons are continuous modes characterized by wave vectors along the axis and their azimuthal symmetry. When a finite metal nanoparticle is placed in the proximity of the nanowire, the discrete plasmon modes of the particle hybridize with the wire plasmon continuum through EM interaction, which is schematically illustrated in Fig. 12. The interaction results in a set of bonding plasmonic virtual states (VS) and antibonding localized states. The VS are primarily composed of long-wavelength wire plasmons and are optically active because of the finite admixture of dipolar nanoparticle plasmons. The nanoparticle serves as a nanoantenna that when polarized by the incident light couples efficiently to wire plasmons of half wavelengths larger than the diameter of the nanoparticle. The wave vector distribution of wire plasmons that make up the VS does not depend on the detailed shape of the nanoparticle but only on its lateral dimension along the wire. The energies of the wire plasmons that participate in the VS are determined by the wire plasmon dispersion relations, which are dependent on the radius of the wire. The composition and spectral properties of the plasmonic VS are thus determined by the nanoparticle and wire diameters rather than by the detailed geometric structure on the nanoparticle. As a result, the EM enhancement is also remarkably insensitive to the detailed geometry. Moreover, the resonant EM enhancement in the nanogap is highest when the external electric field is polarized perpendicular to the nanowire and across the nanogap, because of the strongest EM coupling between the two entities.

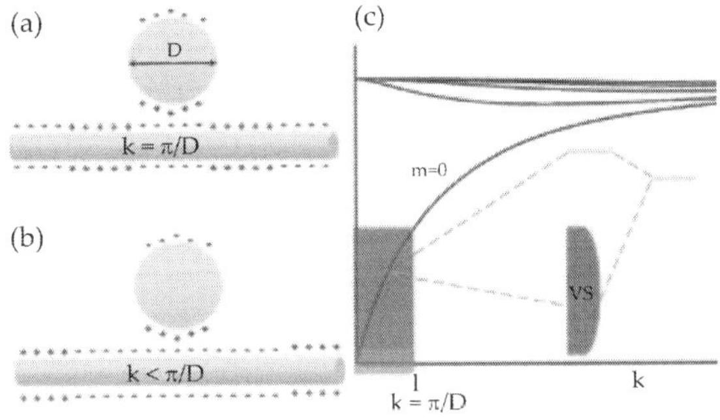


**Figure 12**. Schematic illustration of the coupling of a nanoparticle to the symmetric (m′ = 0) wire plasmon continuum. (a, b) The nanoparticle can couple efficiently to symmetric wire plasmons of half wavelengths larger than the diameter D, (c) Plasmon hybridization in the metal nanoparticle–nanowire system. The red area illustrates the wave vectors of the wire plasmons that couple efficiently to the nanoparticle, and the green area shows their energies. The VS continuum can couple efficiently to light because of the admixture of nanoparticle plasmons with dipolar moments.

The particle–wire geometry also enables remote excitation SERS. The illumination of one end of the nanowire with a laser can launch the SPs in the wire, which propagate to the site of the nanoparticle and excite its plasmons, leading to a strong coupling between the particle and the wire. The high EM enhancement in the nanogap between the nanowire and the nanoparticle renders singlemolecule SERS to be achieved remotely from the laser illumination point. Moreover, when several nanoparticles are attached to one nanowire, multisite remote excitation SERS can be realized in multiple particle–wire nanogaps.

By putting a metal nanoparticle into a nanohole in metal film, a nanogap can be formed between them. The magnitude of the

EM enhancement in the particle–hole nanogap is similar to a nanoparticle dimer. But the volume with large enhancement accessible to molecules is much larger for the particle–hole pair than for the particle dimer. A nanogap can also be created by depositing a nanoparticle onto a metal film with a very thin spacer between them. The large area of the metal film makes it easy to prepare this kind of nanogap. Figure 13 shows the SEM/ transmission electron microscopy (TEM) images of nanoparticle–nanohole and nanoparticle–film structures and the calculated electric field distributions. It can be clearly seen that the hot spots with large electric field enhancement are located within the nanogaps.

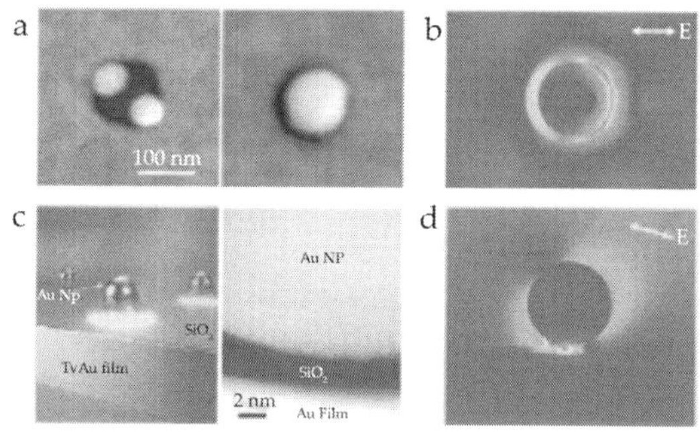


**Figure 13**. (a) SEM images of Au nanoparticle–nanohole structures. (b) Calculated spatial distribution of the electric field for a coupled nanoparticle– nanohole with an excitation wavelength of 633 nm. The diameters of the particle and hole are 100 nm and 120 nm, respectively. The smallest separation between them is 3 nm. (c, left) Schematic diagram of a nanoparticle–film structure; (c, right) high-resolution TEM image of an individual Au nanoparticle on a 2 nm $SiO_2$ spacer, showing its clear separation from the underlying Au film. (d) Calculated spatial distribution of the electric field with an excitation wavelength of 633 nm. The diameter of the nanoparticle is 35 nm, and the thickness of the $SiO_2$ layer is 3 nm.

Nanoparticles can be aligned to form ordered arrays. In these arrays, collective photonic effects arising from the long-range order are superposed onto the plasmon excitation of individual array units, such as nanoparticles or dimers. Theoretical modeling and analysis have shown that photonic effects may further increase the EM enhancement of individual units, when their separations are comparable to the plasmon resonance wavelength. Figure 14 shows the structure of modeled 1D array of dimers composed of nanoshells, which are metal shells supported by dielectric cores. When a whole dimer array is excited by the plane wave, plasmon modes of dimers are excited and consequently long-range dipolar fields are radiated. The dipolar interaction between dimers enhances the effective field experienced by individual dimers compared to the external field, rendering an additional enhancement in the nanogaps of the dimers. Thanks to the $E^4$ law for SERS, the additional enhancement can reach up to 1 order of magnitude, as shown by Fig. 15.

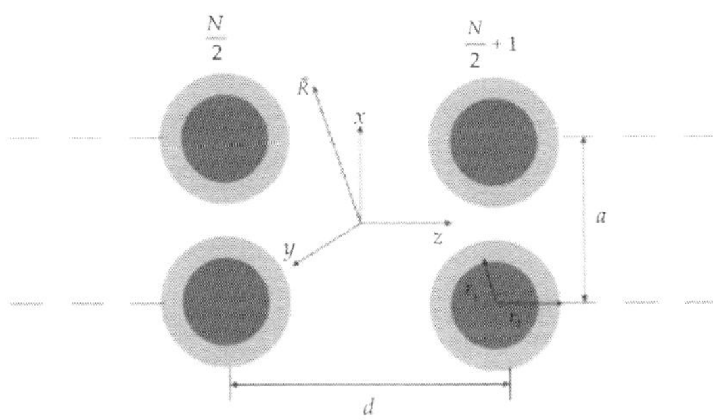


**Figure 14**. Geometrical illustration of an ordered 1D array of metal nanoshell dimers. The metal shell is silver, and the dielectric core is silica.

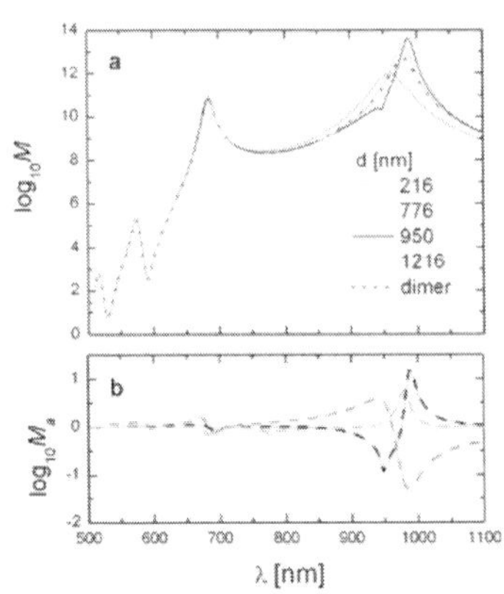


**Figure 15**. (a) EM enhancement versus the wavelength of the incident light at the location halfway between the two centers of a metal nanoshell

dimer in dimer arrays of different interdimer distances d. The inner and outer radii of the nanoshell are 35 nm and 38 nm, respectively. The separation distance between the two nanoshells in a dimer is 1 nm. The dotted line is for an isolated dimer of identical geometry. All the curves are calculated by GMT. (b) The solid lines are the additional enhancements due to long-range photonic effects, given by the subtractions of the dotted line in (a) from the solid lines in (a). The dashed lines are the results from the dipole approximation, which takes account of the interdimer interaction only by dipolar interaction.

Several different types of ordered nanoparticle arrays have been fabricated and used for SERS. In particular, Gunnarsson et al. fabricated highly uniform arrays of nanogaps using electron beam lithography (EBL) to study the interparticle coupling effects for SERS, with the SEM images of the arrays shown in Fig. 16. EBL has the advantage of being able to delicately control the dimensions of the systems and the coupling between nanoparticles. They found that the dependence of SERS intensity on interparticle separation d can be quite well fitted by $I^{SERS} \propto (d + D)^{-2}$ [A(D/d + 1)$^4$ + B where D is the particle length scale and A and B are fitting parameters. Sawai et al. prepared silver dimer arrays using the nanosphere lithography technique and detected SERS spectra from a small number of molecules in the nanogaps of the dimers. The highly uniform and controllable nanogap arrays may be useful for largescale SERS substrates or other optical applications.

**Figure 16.** Scanning electron micrographs of SERS substrates. Examples consist of circular (top left and right), triangular (bottom left), and

square (bottom right) 30 nm thin Ag particles on Si. The predefined particle length scale, defined as the diameter in the case of circular particles and the edge length in the case of triangles or squares, was D = 200 nm, and the predefined interparticle separation distance, defined as the minimum edge-to-edge distance, was d = 100 nm.

## 3.5 MULTIPLE-PARTICLE NANOANTENNAS FOR CONTROLLING POLARIZATION OF SERS EMISSION

The nanogap in aggregated nanoparticles also provides another dimension to tune the properties of SERS by manipulating the polarization of light emitted from molecules trapped in the gaps. To significantly rotate the polarization, one or more nanoparticles must be added to the dimer in order to break the dipolar symmetry of nanoparticle aggregates. The simplest aggregate capable of modulating the polarization is a trimer composed of three nanoparticles, whose asymmetry results in a wavelength-dependent rotation of polarization of the Raman scattering light. Figure 17 shows the rotation of polarization calculated by GMT for the trimer, with the third particle placed at different positions relative to the dimer. To quantify the rotation, the depolarization ratio is used:

$$\rho(\theta) - \frac{I_x(\theta) - I_y(\theta)}{I_x(\theta) + I_y(\theta)}, \tag{6}$$

where $I_x$ and $I_y$ are the components of the emission projected on the x-y coordinate system. Figure 17b shows that the breaking of symmetry may not only rotate the polarization of scattered light but also make it elliptical as characterized by a maximum depolarization ratio $\varrho_{max}$ smaller than 1. The ellipticity of the emission is defined as

$$T = (1 - \rho_{max}) / (1 + \rho_{max}). \tag{7}$$

The ellipticity in Fig. 17b is T = 0.18. The experimentally measured SERS emission from the single molecule in the nanogap of a Ag nanoparticle trimer shows wavelength-dependent polarization rotation and ellipticity, which agrees well with the simulation results. Further simulations show that the polarization rotation is also dependent on the distance between the third particle and the dimer and the size of the third particle.

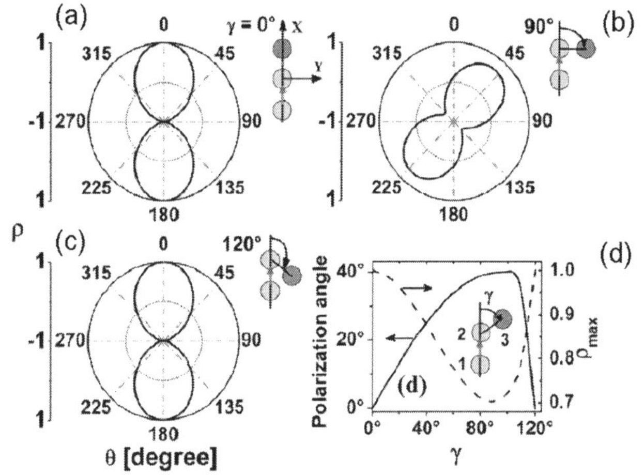


**Figure 17**. Depolarization ratio of the dipole emission from a Ag trimer antenna with (a) linear, (b) right-angle, and (c) equilateral triangle configurations. (d) Polarization angle (left axis) and the maximum depolarization ratio $\varrho_{max}$ (right axis) of the antenna emission as a function of angle $\gamma$. The radii of all three Ag nanoparticles are identical (R = 40 nm). The separations between the first and second nanoparticles and the second and third nanoparticles are kept at 1 nm. The wavelength of the dipole emission is 555 nm. The refractive index of the surrounding medium is $n_s = 1$.

## 3.6 ELECTRONIC COUPLING IN NANOGAPS

When nanoparticles are so close that they nearly touch each other, their electron orbitals may overlap and hybridize, which strongly modifies the electronic structure of the aggregates and thus influences their optical properties. One consequence of the nearly touching limit is a dramatic blue shift of plasmon energy of the aggregates when the interparticle separation decreases to the conductively coupled regime. This effect has been demonstrated by the measured optical transmission spectra of an array of gold nanodisk dimers, as shown in Fig. 18. The valleys of spectra correspond to the LSP energy of dimers (long-range interdimer interaction has little effect on the LSP energy) because incident light with those frequencies is absorbed and converted into plasmon excitation. When the electron orbitals of individual nanodisks are distinctly separated from each other and the electronic coupling between disks is weak, the LSP energy red-shifts with the decreasing gap size as a result of plasmon hybridization. But at the nearly touching regime, the LSP energy suddenly displays a significant blue shift due to strong electronic coupling between nanodisks and a substantially modified electronic structure of the whole dimer.

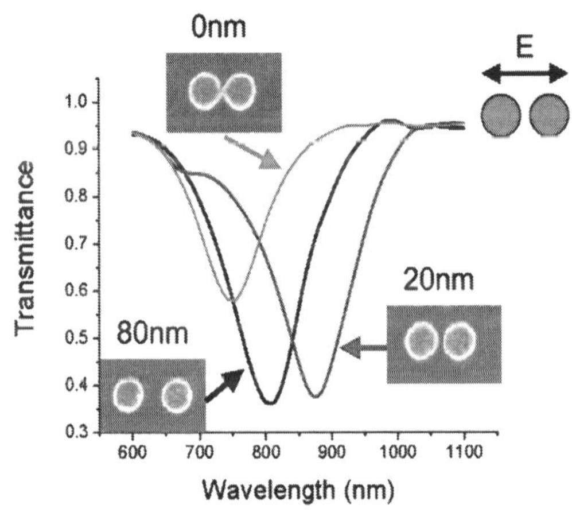


**Figure 18**. Polarized transmission spectra in periodic arrays of pairwise interacting gold nanoparticles. The lattice constant is 800 nm in the parallel direction to the pair axis and 400 nm in the perpendicular direction, and the dot height is 30 nm. The separation distance between the two particles of a pair is labeled above the SEM images.

Another consequence of the strong electronic coupling is that electrons may tunnel across the dimer junction, significantly modifying the optical response and reducing the EM enhancement relative to the classical predictions. Quantum mechanical calculations indicate that this tunneling effect is important for a gap size smaller than ~1 nm, which may still hold small molecules. For larger gap sizes, the results of quantum mechanical and classical calculations gradually merge, further confirming the validity of applying classical EM theory to aggregates with wide enough gaps. Using time-dependent local density approximation (TDLDA), Zuloaga et al. studied EM enhancement in the nanogap between two electronically coupled jellium nanospheres and explicitly demonstrated their difference from results obtained by classical theory. As shown in Fig. 19, three distinct regimes of interaction between nanoparticles are identified: the classical regime (Figs. 19a-i and 19b-A), where the gap separation d is larger than ~1 nm and interparticle interaction can be well characterized by plasmon hybridization; the crossover regime (Figs. 19a-ii, 19b-B, and 19b-C), where electrons start to tunnel through a narrow barrier, reducing the EM coupling between the two particles and consequently EM enhancement in the nanogap; and the conductive regime (Figs. 19a-iii and 19b-D), where the Fermi level of the system lies above the electron potential barrier separating the two particles and the conductance of the gap is large. The enhancement in the crossover and conductive (touching) regimes clearly differentiates quantum mechanical and classical predictions (Fig. 19b).

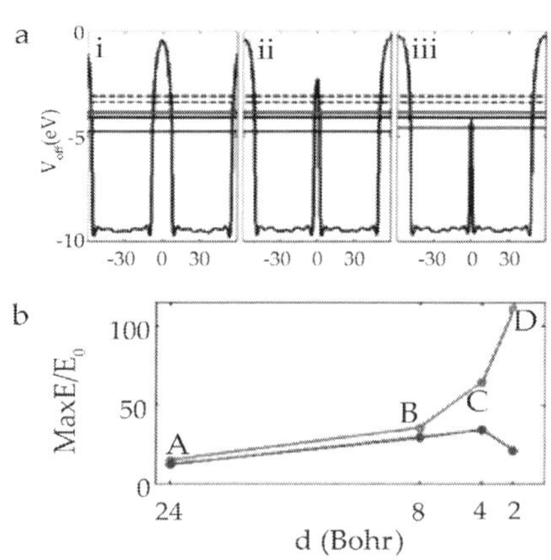


**Figure 19**. (a) Self-consistent effective potential along the interparticle axis of the dimer for a nanosphere of radius R = 24b and separation d =

16b (i), 5b (ii), and 2b (iii). Here b is Bohr's radius, which is 0.0529 nm. The top horizontal, solid lines mark the Fermi level. (b) Comparison of the maximum electric field enhancements calculated using plasmon hybridization (upper) and TDLDA (lower) for R = 24b dimers of different separation d.

## 3.7 PROBING EM ENHANCEMENT VIA SERS

Because of the small scale of hot spots, using normal sensors to measure EM field distribution is generally not feasible. Yet the EM effect should enable molecules to effectively probe the field distribution by measuring their SERS signal, as long as the chemical effect can be suppressed. Wang et al. combined simulations and experiments to determine the field distribution by studying the SERS of rhodamine 6G (R6G) molecules as they diffuse into Ag@$SiO_2$ coreshell nanoparticles. The Ag@$SiO_2$ nanoparticle is composed of a $SiO_2$ shell supported by a Ag core. When R6G diffuses in the confined $SiO_2$ nanoporous structure and does not chemically bind to Ag, SERS should purely originate from EM enhancement, thanks to LSPs of the Ag core. Mie theory is used to simulate the field distribution in the $SiO_2$ shell, and diffusion theory is applied to calculate the evolution of the R6G concentration in the shell. The concentration of R6G within the $SiO_2$ shell C(r, t) is obtained by solving the equation

$$\frac{\partial C(r,t)}{\partial t} = D\frac{1}{r^2}\frac{\partial}{\partial r}\left[r^2\frac{\partial C(r,t)}{\partial r}\right], \qquad (8)$$

where D is the effective diffusion coefficient, t is the time, and r is the radial coordinate. D is the only adjustable parameter used in calculations. The average SERS enhancement factor $G_{ave}$ is obtained by integrating all Raman scattering contributions from the volume of the spherical shell (v):

$$G_{ave}(t) = \int_V G_{loc}(r)C(r,t)dV, \qquad (9)$$

where $G_{loc}$ is the local enhancement at spot r in the shell. Theoretical values of average enhancement $G_{ave}(t)$ in the $SiO_2$ shell are thus obtained and compared to experimental measurement. The good agreement between theoretical and experimental results in Fig. 20a is a solid proof that the calculated field distribution is an excellent simulation of the true distribution. Figure 20b shows the EM field distribution quantified by local enhancement.

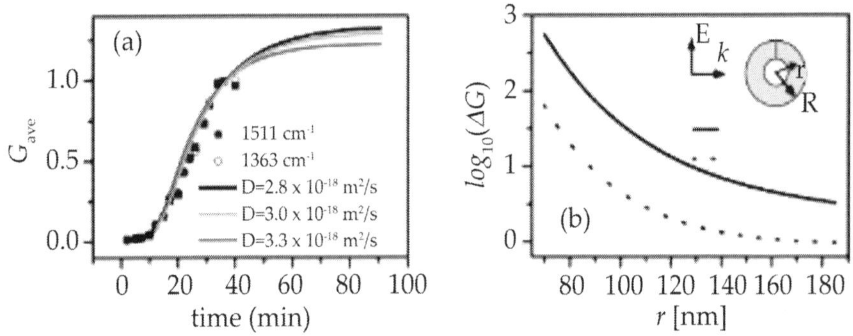


**Figure 20.** (a) Comparison between experimentally observed (solid squares and open circles) and theoretically predicted (lines) SERS enhancement factor, $G_{ave}$, using spectral data at 1511 or 1363 $cm^{-1}$. All data were normalized to the saturation value collected at 40 min. The estimated effective diffusion coefficient, D, is $2.8 \times 10^{-18}$ to $3.3 \times 10^{-18}$ $m^2/s$. (b) Local EM intensity enhancement ($\Delta G$) along the transect of the silica shell shown in the inset (solid line) and the average $\Delta G$ over the surface at radius r (dashed line).

By measuring near-field Raman scattering and near-field twophoton-induced photoluminescence using an apertured scanning near-field optical microscope, Imura et al. demonstrated that the large EM enhancement in a dimer of gold nanoparticles is localized in the nanogap between the two particles when the laser light is polarized parallel to the interparticle axis. The hot spots in the nanogaps can be experimentally mapped by using the superresolution imaging technique. Willets et al. mapped the singlemolecule SERS hot spot by point spread function fitting, and a resolution of the order of ~5 nm was achieved. The superresolution optical images reveal the size and field distribution of the hot spots in nanogaps of aggregated nanoparticles. These results provide direct experimental evidence for the large EM enhancement in the nanogaps and its dominant contribution to single-molecule SERS.

## REFERENCES


1. A. M. Michaels, J. Jiang, and L. Brus, J. Phys. Chem. B, 104, 11965–11971 (2000).
2. A. Taove, and S. C. Hagness, Computational Electrodynamics: The FiniteDifference Time-Domain Method, 3rd ed. (Artech House, 2005).
3. C. M. Ruan, G. Eres, W. Wang, Z. Y. Zhang, and B. H. Gu, Langmuir, 23, 5757–5760 (2007).
4. C. Troparevsky, K. Zhao, D. Xiao, A. G. Eguiluz, and Z. Y. Zhang, Phys. Rev. B, 82, 045413 (2010).
5. E. M. van Schrojenstein Lantman, T. Deckert-Gaudig, A. J. G. Mank, V. Deckert, and B. M. Weckhuysen, Nat. Nanotechnol., 7, 583–586 (2012).
6. F. Hao and P. Nordlander, Appl. Phys. Lett., 89, 103101 (2006).
7. Grigorenko, S. Haas, and A. F. J. Levi, Phys. Rev. Lett., 97, 036806 (2006).
8. Gunnarsson, E. J. Bjerneld, H. X. Xu, S. Petronis, B. Kasemo, and M. Käll, Appl. Phys. Lett., 78, 802–804 (2001).
9. H. Wei and H. X. Xu, Nanoscale, 5, 10794–10805 (2013).
10. H. Wei, F. Hao, Y. Z. Huang, W. Z. Wang, P. Nordlander, and H. X. Xu, Nano Lett., 8, 2497–2502 (2008).
11. H. Wei, U. Hakanson, Z. L. Yang, F. Hook, and H. X. Xu, Small, 4, 1296–1300 (2008).
12. H. X. Xu and M. Käll, ChemPhysChem, 4, 1001–1005 (2003).
13. H. X. Xu, and M. Käll, Estimating SERS properties of silver-particle aggregates through generalized Mie theory. K. Kneipp, M. Moskovits and H. Kneipp, eds., Surface-Enhanced Raman Scattering: Physics and Applications (Springer, 2006), pp. 87–104.
14. H. X. Xu, Appl. Phys. Lett., 85, 5980–5982 (2004).
15. H. X. Xu, J. Aizpurua, M. Käll, and P. Apell, Phys. Rev. E, 62, 4318–4324 (2000).
16. J. A. Hutchison, S. P. Centeno, H. Odaka, H. Fukumura, J.

Hofkens, and H. Uji-i, Nano Lett., 9, 995–1001 (2009).
17. J. M. Pitarke, V. M. Silkin, E. V. Chulkov, and P. M. Echenique, Rep. Prog. Phys., 70, 1–87 (2007).
18. J. N. Chen, W. S. Yang, K. Dick, K. Deppert, H. Q. Xu, L. Samuelson, and H. X. Xu, Appl. Phys. Lett., 92, 093110 (2008).
19. J. Steidtner and B. Pettinger, Phys. Rev. Lett., 100, 236101 (2008).
20. K. Imura, H. Okamoto, M. K. Hossain, and M. Kitajima, Nano Lett., 6, 2173–2176 (2006).
21. K. J. Savage, M. M. Hawkeye, R. Esteban, A. G. Borisov, J. Aizpurua, and J. J. Baumberg, Nature, 491, 574–577 (2012).
22. K. Zhao, H. X. Xu, B. H. Gu, and Z. Y. Zhang, J. Chem. Phys., 125, 081102 (2006).
23. L. D. Qin, S. L. Zou, C. Xue, A. Atkinson, G. C. Schatz, and C. A. Mirkin, Proc. Natl. Acad. Sci. USA, 103, 13300–13303 (2006).
24. L. Mao, Z. P. Li, B. Wu, and H. X. Xu, Appl. Phys. Lett., 94, 243102 (2009).
25. M. L. Weber and K. A. Willets, J. Phys. Chem. Lett., 2, 1766–1770 (2011).
26. M. McMahon, S. K. Gray, and G. C. Schatz, Phys. Rev. Lett., 103, 097403 (2009).
27. M. P. Marder, Condensed Matter Physics, 1st ed. (Wiley, New York, 2000).
28. N. Hayazawa, Y. Inouye, Z. Sekkat, and S. Kawata, Opt. Commun., 183, 333–336 (2000).
29. P. G. Etchegoin, E. C. Le Ru, and M. Meyer, Phys. Today, 61, 13–14 (2008).
30. P. Nordlander, C. Oubre, E. Prodan, K. Li, and M. I. Stockman, Nano Lett., 4, 899–903 (2004).
31. R. Esteban, A. G. Borisov, P. Nordlander, and J. Aizpurua, Nat. Commun., 3, 825 (2012).
32. R. Zhang, Y. Zhang, Z. C. Dong, S. Jiang, C. Zhang, L. G. Chen, L. Zhang, Y. Liao, J. Aizpurua, Y. Luo, J. L. Yang, and J.

G. Hou, Nature, 498, 82–86 (2013).
33. S. L. Zou and G. C. Schatz, Chem. Phys. Lett., 403, 62–67 (2005).
34. S. Mubeen, S. P. Zhang, N. Kim, S. Lee, S. Kramer, H. X. Xu, and M. Moskovits, Nano Lett., 12, 2088–2094 (2012).
35. T. Atay, J. H. Song, and A. V. Nurmikko, Nano Lett., 4, 1627–1631 (2004).
36. T. Shegai, Z. P. Li, T. Dadosh, Z. Y. Zhang, H. X. Xu, and G. Haran, Proc. Natl. Acad. Sci. USA, 105, 16448–16453 (2008).
37. T. Sun, Z. L. Zhang, H. R. Zheng, and H. X. Xu, Sci. Rep., 2, 647 (2012).
38. W. L. Barnes, A. Dereux, and T. W. Ebbesen, Nature, 424, 824–830 (2003).
39. W. Wang, Z. P. Li, B. H. Gu, Z. Y. Zhang, and H. X. Xu, ACS Nano, 3, 3493–3496 (2009).
40. Y. R. Fang, H. Wei, F. Hao, P. Nordlander, and H. X. Xu, Nano Lett., 9, 2049–2053 (2009).
41. Y. Sawai, B. Takimoto, H. Nabika, K. Ajito, and K. Murakoshi, J. Am. Chem. Soc., 129, 1658–1662 (2007).
42. Z. L. Yang, J. Aizpurua, and H. X. Xu, J. Raman Spectrosc., 40, 1343–1348 (2009).
43. Z. Liu, S. Y. Ding, Z. B. Chen, X. Wang, J. H. Tian, J. R. Anema, X. S. Zhou, D. Y. Wu, B. W. Mao, X. Xu, B. Ren, and Z. Q. Tian, Nat. Commun., 2, 305 (2011).
44. Z. P. Li, T. Shegai, G. Haran, and H. X. Xu, ACS Nano, 3, 637–642 (2009).
45. Z. Q. Tian, B. Ren, and D. Y. Wu, J. Phys. Chem. B, 106, 9463–9483 (2002).
46. Zuloaga, E. Prodan, and P. Nordlander, Nano Lett., 9, 887–891 (2009).

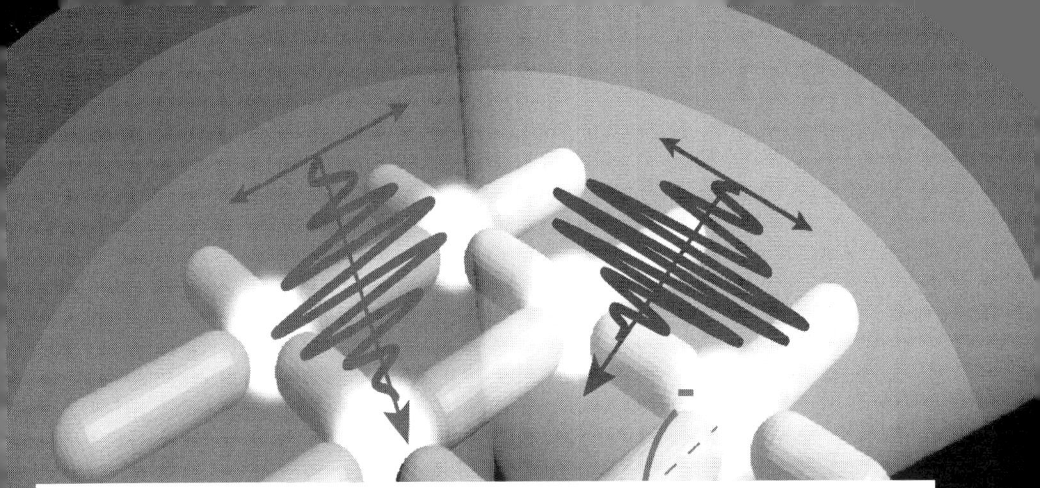

CHAPTER 4

# PLASMONIC ANTENNAS

## INTRODUCTION

Antennas were first developed a century ago as electrical devices to convert electric power into radio waves and vice versa. They play an essential role in modern wireless communication for radiating and receiving radio waves. Plasmonic antennas as an analogue at the nanoscale are of great interest due to the unique ability of metallic nanostructures to manipulate the absorption and emission at optical frequency, such as focusing light to volumes far beyond the diffraction limit, enhancing the Raman scattering and fluorescence of molecules and quantum emitters, manipulating their emission polarization, and modifying their lifetime. Mediated by the surface plasmon resonance (SPR) of metallic nanostructures, the plasmonic antennas can effectively convert the free propagating light into the nanoscaleenhanced near field and vice versa, and a localized excitation can be coupled out as far-field radiation. The efficiency of a plasmonic antenna depends on its shape, material, dimension, assembly, and operation frequency. So far, various plasmonic antennas have been developed experimentally and theoretically to tune the SPR, such as nanodisks, triangles, and flowers, as well as surface-plasmon-coupled antennas, including dimers, bowties, and trimers. These single and coupled plasmonic antennas have been investigated thoroughly by various far-field and near-field techniques, such as far-field scattering spectroscopy, electron energy loss spectroscopy, nonlinear two-photon excited luminescence, and nearfield scanning optical microscopy. Abundant applications have been found in surface-enhanced Raman scattering (SERS), optical manipulation, biosensing, and integrated photonic devices.

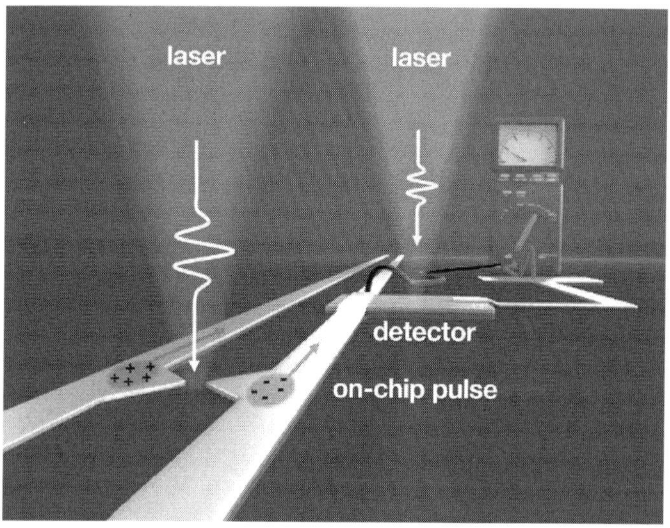


## 4.1 SINGLE PLASMONIC ANTENNAS

SPR is a kind of collective oscillation of a free electron cloud on the surface of metallic nanostructures, induced by incident light. It can enlarge effectively the cross section of nanostructures interacting with light. The absorption and scattering spectra of silver nanostructures of different shapes but with similar dimensions (both the side of the cube and the diameter of the ring are 50 nm) are shown in Fig. 1a, which presents a prominent resonant feature within the visible range due to the excitation of dipolar SPR. The cubic one has a broad peak at around 430 nm, whereas for the ring case, the peak is shifted to 568 nm. It demonstrates that the shape can greatly affect the frequency of SPR excited in metallic nanoparticles, which makes them good candidates for nanoantennas at different optical frequencies. Experimentally, the dependence of SPR on shape can be demonstrated by performing the single-particle spectroscopy. Figure 1b shows the continuous tunability of SPR from visible to infrared regions in anisotropic nanostructures by varying the aspect ratio.

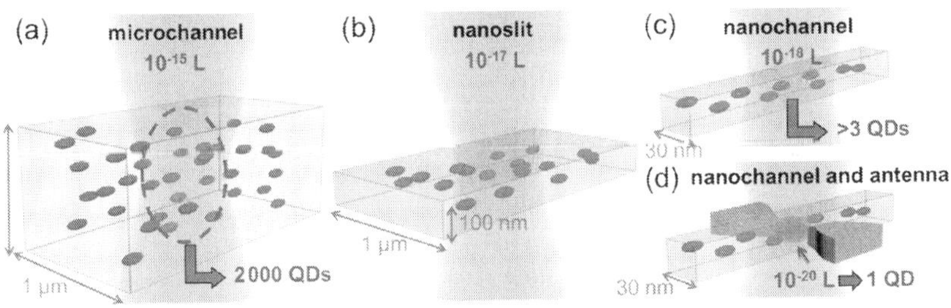


Here, the widths of the particles are kept at around 55 nm, with the lengths varying from 100 to 200 nm. Two prominent resonance peaks can be observed in each single-particle dark-field scattering spectrum. One is located at around 430 nm and shows little dependence on the elongation of the length. This peak corresponds to the surface plasmon (SP) mode with charge oscillation along the short axis of the particle, that is, the transverse dipole plasmon mode. The other one red-shifts from 620 to 850 nm with the increase in the aspect ratio of particles. This peak is originated from the SP oscillation along the long axis of the particle, that is, the longitudinal dipole plasmon mode. Hence, it presents a great dependence on the particle length. Not only the tunable far-field radiation but also a huge local field enhancement can be obtained in the vicinity of plasmonic antennas under resonant excitation.

In Fig. 1c, the calculated near field around a gold nanobar shows the excitation of the first three longitudinal plasmon modes supported by a 400 nm long silver nanobar with a 50 × 100 nm$^2$ cross section under the free-space excitation wavelengths $\lambda_{eff}$

= 1375, 770, and 630 nm. Here, the longitudinal resonances of a nanobar can be understood as an oscillation in the Fabry–Pérot cavity. The approximate relationship between the rod length $L$ and the effective resonance wavelength $\lambda_{\text{eff}}$ of longitudinal modes can be written as

$$L = \frac{n}{2}\lambda_{\text{eff}},$$

where $n$ is the integer number indicating the order of longitudinal plasmon resonance. The instantaneous charge configurations for the first three orders of longitudinal plasmon resonances are shown at the bottom of Fig. 1c.

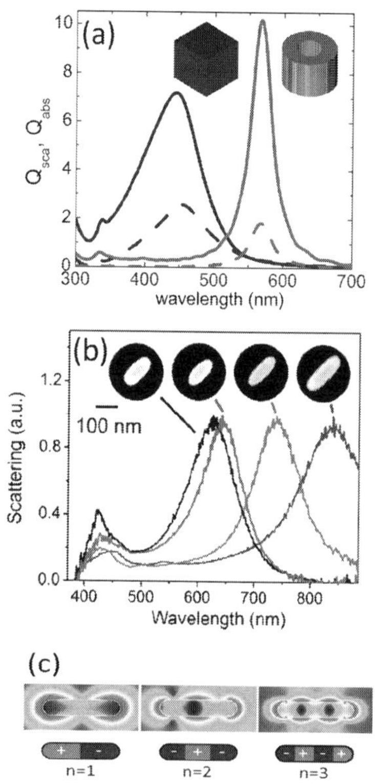


**Figure 1:** (a) Calculated absorption (solid lines) and scattering (dashed lines) efficiencies for a cubic-shaped and a ring-shaped silver nanoparticle. (b) Experimentally measured dark-field scattering spectra of single silver nanobars with different aspect ratios. (c) Electric field intensity enhancement corresponding to the first three orders of longitudinal plasmon resonances.

The large near-field enhancement determines the ability of antennas to focus the energy into a subdiffraction-limited volume. Experimentally describing the near-field distribution is essential for the research and practical applications of plasmonic antennas. Nonlinear microscopy is a promising technique to visualize the highly confined electric field, because the higher power dependence of nonlinear responses on the electric field makes it preferentially sensitive to the most intense field close to the metal. For instance, the two-photon-induced luminescence (TPL) through interband transition of gold can provide an effective method to characterize the near-field distribution. The top panel of Fig. 2a shows an example of a TPL image from a single gold antenna under longitudinal excitation. Excited by the intense pulsed radiation at 730 nm, not only the longitudinal plasmon resonances, but also a TPL signal can be observed. The distribution of the TPL intensity around the antenna can be further mapped by scanning the antenna with respect to the illumination spot. Interestingly, it is found that the TPL signal is concentrated at the extremities of the gold bar antenna. The scanning electron microscopy (SEM) image of this nanobar is shown in the bottom panel of Fig. 2a. The superimposed black lines plot the TPL signal along the symmetry axis of the antennas. Owing to the quadratic dependence of the TPL signal on excitation intensity $I^2$, the observed TPL map should well reflect the distribution of the local field to the fourth power $|E|^4$. The top panel of Figure 2b shows the calculated distribution of $|E|^4$ for resonant plasmon modes. Additionally, it is known that the optical resolution in experiments is about 200 nm. So the calculated electric field map is further convoluted by a 2D Gaussian profile with a 200 nm fullwidth at half-maximum (FWHM). The resulting field distribution is fully dominated by the strong intensity at the extremities, as shown in the bottom panel of Fig. 2b. By comparison, it is found that the TPL image can well reflect the local field distributions of resonant plasmon modes, which provides a useful technique for the highly confined electric field visualization around the gold nanoantennas.

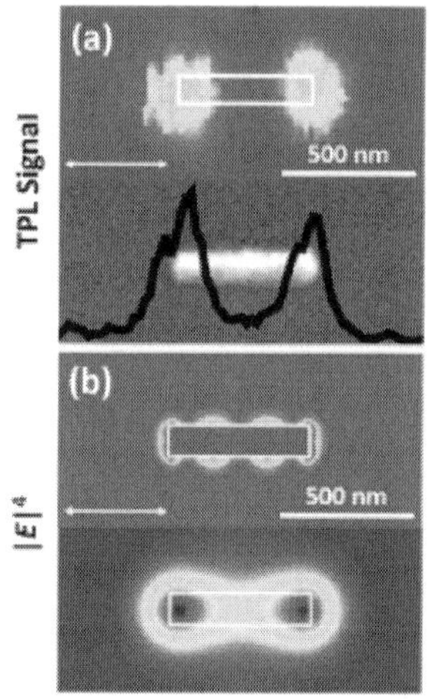


**Figure 2:** (a, top) TPL scans for a single 500 nm long gold bar under pulsed laser excitation at 730 nm. (a, bottom) the SEM image. (b, top) Calculated $|E|^4$ distribution for resonant plasmon modes under longitudinal polarization at 710 nm excitation; (b, bottom) convoluted $|E|^4$ distribution by a 2D Gaussian profile with 200 nm FWHM.

Besides the TPL image, the near-field distribution can also be visualized by the scanning near-field optical microscope (SNOM). As mentioned before, the shape of the antenna influences both farfield and near-field properties. Figure 3a shows the darkfield scattering spectrum of a triangular nanoantenna with edge length 200 nm. Three typical SEM images of triangle antennas of that size are shown in the insets. The scale bar is 200 nm. Two peaks dominate the spectrum. One is at ~670 nm and the other at ~430 nm, corresponding to the dipole resonance and quadrupole dark mode, respectively. Here, we should note that the quadrupole mode becomes radiative through hybridization with the dipole mode. Figure 3b shows the size effect on the resonant frequencies of dipolar and quadrupolar plasmon modes. The dipolar and quadrupolar resonances progressively shift to lower energies with the increase of the edge length.

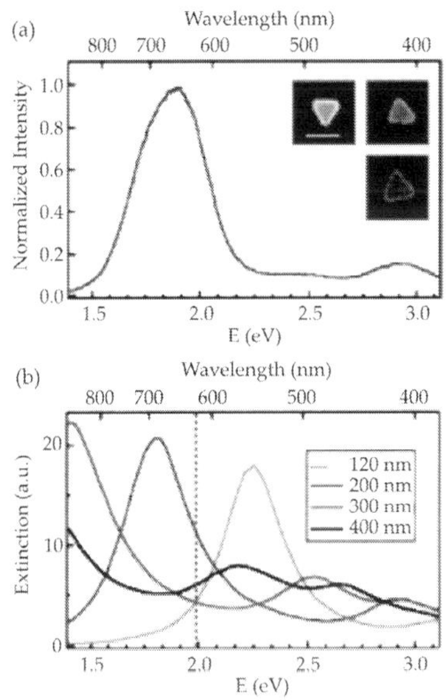


**Figure 3:** (a) The normalized scattering spectrum of an isolated silver nanotriangle with edge length 200 nm. (b) The scattering spectra of nanotriangles with edge lengths 120, 200, 300, and 400 nm.

The size dependence of these two modes is then visualized by an SNOM under an excitation of 633 nm, indicated by the red dashed line in Fig. 3b. Figure 4a shows the SNOM image of a silver nanotriangle with edge length ~120 nm. The atomic force microscopy (AFM) topography is shown in the inset. Under an s-polarized incidence of 633 nm, the in-plane dipolar resonance can be excited, as discussed in Fig. 3b. The observed near-field pattern exhibits prominent dipolar resonance features, with the strongest local field at two tips and extended spatial distribution beyond the particle boundaries, which is consistent with the field distribution of a typical dipole. The numerical simulation in Fig. 4b using the discrete dipole approximation (DDA) also gives the dipolar resonance features (the plus and minus signs indicate their relative phases), which corresponds well to the SNOM results. For the triangle with a larger edge length ~450 nm, the resonance is dominated by the quadrupole plasmon mode under the 633 nm excitation, as indicated in Fig. 3b. Figure 4c shows the corresponding SNOM result, and the AFM topography image is shown in the inset. Under the s-polarized incidence, the SNOM image shows a typical quadrupole response featured by four local-fieldenhanced regions. It gives rise to the largest near-field enhancement at the lower two adjacent edges. And the other two weaker charge centers of the quadrupole can be identified at the two tips of the top edge. The DDA simulation under the experimental conditions also shows good accordance with the experiments, as shown in Fig. 4d. These results demonstrate that an SNOM can also well visualize the local electric field distributions of resonant plasmon modes under a certain excitation.

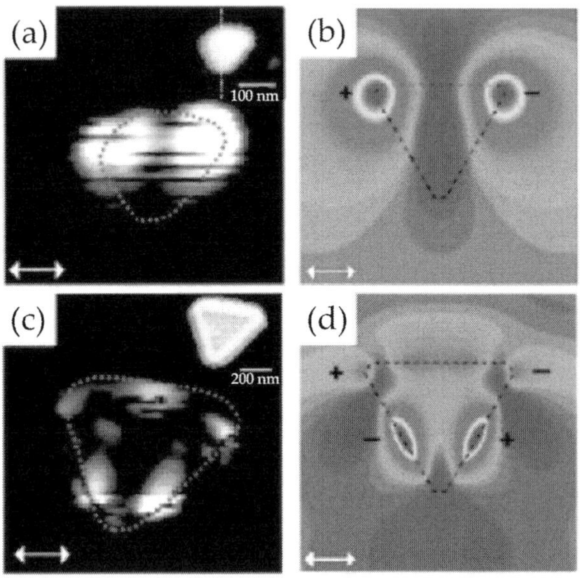


**Figure 4:** (a, b) Experimental and theoretical distributions of the local electric field for a nanotriangle (edge length ~120 nm) supporting a dipole plasmon resonance. Panels (c) and (d) are identical to panels (a) and (b) but for a larger nanotriangle (edge length ~450 nm) under quadrupole excitation.

## 4.2 OPTICAL ANTENNAS

While radio antennas were developed as solutions to a communication problem, the invention of optical antennas was motivated by microscopy. In analogy to its radio wave and microwave counterparts, we define the optical antenna as a device designed to efficiently convert free-propagating optical radiation to localized energy, and vice versa. In the context of microscopy, an optical antenna effectively replaces a conventional focusing lens or objective, concentrating external laser radiation to dimensions smaller than the diffraction limit.

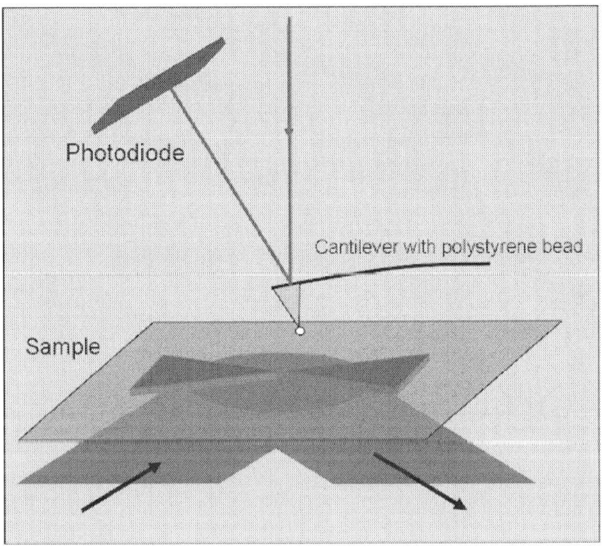


"The particle serves as an antenna that receives an incoming electromagnetic field," making him the first to mention explicitly the analogy of local microscopic light sources to classical antennas—a concept that has since been thoroughly explored. The invention of scanning tunneling microscopy and the discovery of surface enhanced Raman scattering (SERS) most likely inspired Wessel's idea. The quest for an understanding of SERS gave rise to many theoretical studies aimed at predicting the electromagnetic field enhancement near laser-irradiated metal particles and clusters. This era can be considered the first phase of what is now called nanoplasmonics.

Instead of a solid metal particle, they used a gold-coated polystyrene particle as a local light source, a structure that was later extensively developed and is now called a gold nanoshell. A thin metal film with 320 nm holes and demonstrated a spatial resolution of ~

50 nm. Their results provide the first experimental evidence that near-field scanning optical microscopy as introduced by Synge is feasible. Ten years later, laser-irradiated metal tips were proposed as optical antenna probes for near-field microscopy and optical trapping, and since then various antenna geometries have been studied (rods, bowties, etc.)

The optical antenna concept also has strong parallels with the development of so-called whisker diodes in the 1960s. An antenna attached to a metal-to-metal point contact was used in 1968 by Ali Javan and co-workers for frequency mixing of IR radiation.

It was shown that the rectification efficiency of these whisker diodes could be increased by suitably "kinking," or bending the wire antenna. The length L (tip to kink) had to be adjusted in relation to the wavelength and angle of incidence, and the strongest response was obtained for the fundamental resonance of $L \approx \lambda/2.7$. These experiments were performed at IR wavelengths, at which metals are good conductors. IR antenna fabrication has expanded considerably in the past few decades, notably including work by Glen Boreman, who since the late 1990s has fabricated many types for use in IR detectors, bolometers, and nanophotonics. Figure 5 catalogs several examples of these antennas. In search of a near-field optical probe with a higher efficiency than tapered fibers. A microwave-scale model of a bowtie antenna at the end of a waveguide and demonstrated confinement of a tenth of a wavelength $(\lambda)$ with 30% transmission efficiency.

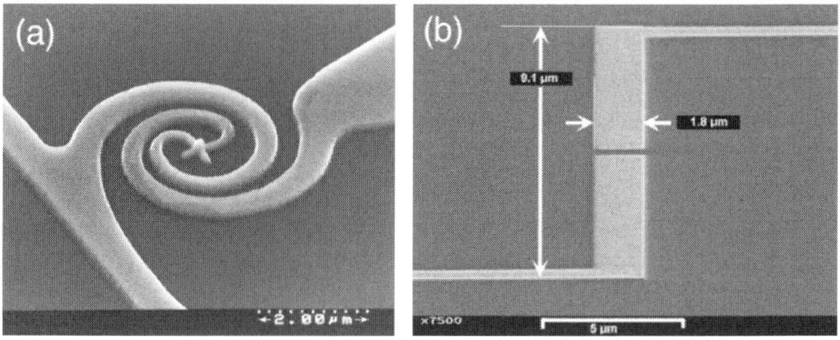

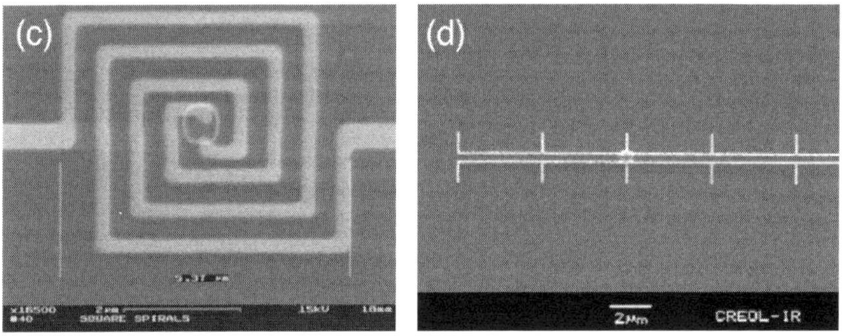


**Figure 5**: catalogs several examples of these antennas

Examples of IR optical antennas fabricated by Boreman and co-workers: (a) asymmetric spiral antenna, (b) microstrip dipole antenna, (c) square spiral antenna, (d) phased-array antenna

## 4.2.1 Physical Properties of Optical Antennas

Optical antennas are strongly analogous to their RF and microwave counterparts, but there are crucial differences in their physical properties and scaling behavior. Most of these differences arise because metals are not perfect conductors at optical frequencies, but are instead strongly correlated plasmas described as a free electron gas. Optical antennas are also not typically driven with galvanic transmission lines—localized oscillators are instead brought close to the feed point of the antennas, and electronic oscillations are driven capacitively. Moreover, optical antennas can take various unusual forms (tips, nanoparticles, etc.) and their properties may be strongly shape and material dependent owing to surface plasmon resonances.

The general problem statement of optical antenna theory is illustrated in Fig. 6. A receiver or transmitter interacts with free optical radiation via an optical antenna. The receiver or transmitter is ideally an elemental quantum absorber or emitter, such as an atom, ion, molecule, quantum dot, or defect center in a solid.

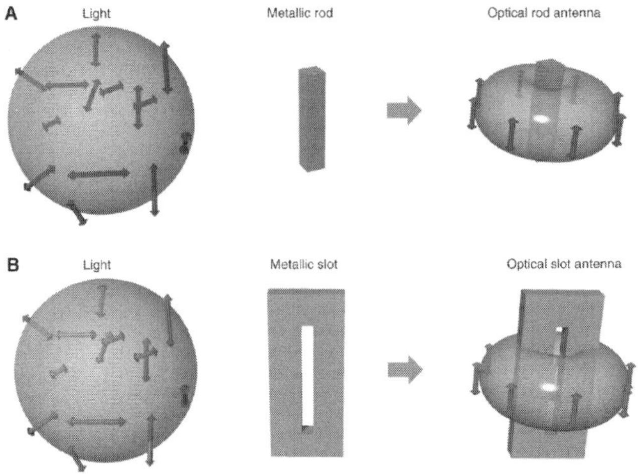


The antenna enhances the interaction between the emitter or absorber and the radiation field. It therefore provides the prospect of controlling the light–matter interaction on the level of a single quantum system. The presence of the antenna modifies the properties of the receiver/transmitter, such as its transition rates and, in the case of a strong interaction, even the energy-level structure. Likewise, the antenna properties depend on those of the receiver–transmitter, and it becomes evident that the two must be regarded as a coupled system. In this section, we will address these issues and attempt to express them in terms of established antenna terminology.

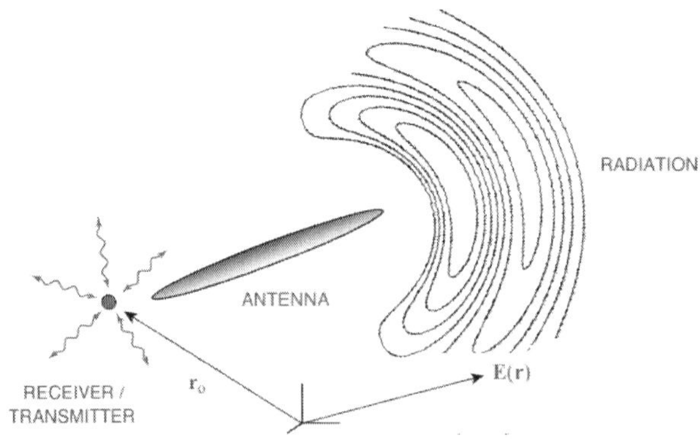


**Figure 6**: Problem statement of optical antenna.

### *Local Density of Electromagnetic States*

Arguably, one of the most important quantities in a discussion of antennas is the impedance, defined in circuit theory in terms of source current I and voltage V as Z=V/I. This definition assumes that the source is connected to the antenna via a current-carrying transmission line. But optical antennas are typically fed by localized light emitters, not by real currents. Thus, the definition of antenna input impedance needs some adjustments. A viable alternate definition involves the local density of electromagnetic states (LDOS), which can be expressed in terms of the Green's function tensor $\overleftrightarrow{G}$ and which accounts for the energy dissipation of a dipole in an arbitrary inhomogeneous environment.

Single emitters such as atoms and molecules are inherently quantum objects that, strictly speaking, demand a quantum mechanical treatment. However, a two-level system that mostly resides in the ground state (perturbation limit) can be represented by a classical dipole. Therefore, we begin the discussion with the quantum mechanical description of a two-level atom and then establish the link to the classical representation.

The total decay rate of a two-level quantum emitter located at $r_o$ and weakly coupled to the antenna can be represented by Fermi's golden rule as

$$\Gamma = \frac{\pi\omega}{3\hbar\varepsilon_o}|\langle g|\hat{\mathbf{p}}|e\rangle|^2 \rho_{\mathbf{p}}(\mathbf{r}_o,\omega), \tag{1}$$

where $\langle g|\hat{\mathbf{p}}|e\rangle$ is the transition dipole moment between the emitter's excited state $|e\rangle$ and ground state $|g\rangle$, $\omega$ is the transition frequency, and $\rho_{\mathrm{p}}$ denotes the partial density of electromagnetic states (LDOS). The latter can be expressed in terms of the system's dyadic Green's function $\overleftrightarrow{G}$ as

$$\rho_{\mathbf{p}}(\mathbf{r}_o,\omega) = \frac{6\omega}{\pi c^2}[\mathbf{n}_p \cdot \mathrm{Im}\{\overleftrightarrow{G}(\mathbf{r}_o,\mathbf{r}_o,\omega)\} \cdot \mathbf{n}_p], \tag{2}$$

where $n_p$ is a unit vector pointing in direction of p. The Green's function used in Eq. (2) is indirectly defined by the electric field E at the observation point r generated by a dipole p located at $r_o$,

$$\mathbf{E}(\mathbf{r}) = \frac{1}{\varepsilon_o}\frac{\omega^2}{c^2}\overleftrightarrow{G}(\mathbf{r},\mathbf{r}_o,\omega)\mathbf{p}. \tag{3}$$

Notice that the Green's function in Eq. (2) is evaluated at $r_o$, which is the position of the emitter itself. This reflects the fact that the decay from the excited state happens in response to the emitter's own field.

The total LDOS $(\rho)$ is obtained by assuming that the quantum emitter has no preferred dipole axis (atom). Averaging Eq. (2) over different dipole orientations leads to

$$\rho(\mathbf{r}_o,\omega) = \langle \rho_{\mathbf{p}}(\mathbf{r}_o,\omega)\rangle = \frac{2\omega}{\pi c^2}\,\mathrm{Im}\{Tr[\overleftrightarrow{G}(\mathbf{r}_o,\mathbf{r}_o,\omega)]\}, \tag{4}$$

where Tr denotes the trace. Thus, the excited state lifetime $\tau = 1/\Gamma$ of the quantum emitter is determined by the Green's function $\overleftrightarrow{G}$ of the system in which the emitter is embedded. The LDOS therefore accounts for the presence of the antenna and is a measure of its properties. In free space, i.e., in the absence of the antenna, we obtain $\rho_{\mathbf{p}} = \omega^2/(\pi^2 c^3)$ and $\Gamma_o = \omega^3|\langle g|\hat{\mathbf{p}}|e\rangle|^2/(3\pi\varepsilon_o\hbar c^3)$. The observation that atomic decay rates are dependent on the local environment goes back to Purcell's analysis in 1946 and has since been measured for various systems, such as molecules near interfaces or atoms in cavities. The origin for the modification of atomic decay rates is the interaction of the atom with its own secondary field—the field that arrives back at the atom's location after being scattered in the local environment. This back-action also influences the energy states and transition frequencies, but the effect is generally much smaller than the modification of transition rates.

### *Power Dissipation and Antenna Impedance*

The Green's function introduced in the previous section corresponds to the electric field of a classical dipole oscillating at frequency $\omega$. We likewise represent the quantum emitter by a classical dipole p, a pointlike source current located at $r_o$. According to Poynting's theorem the power dissipated by a timeharmonic system is

$$P = \frac{1}{2}\int_V \mathrm{Re}\{\mathbf{j}^* \cdot \mathbf{E}\}dV, \tag{5}$$

where V is the source volume, j the current density, and E the electric field. The current density j can be expanded in a Taylor series around some origin $r_o$ and to lowest approximation can be written as

$$\mathbf{j}(\mathbf{r}) = -i\omega\mathbf{p}\,\delta[\mathbf{r}-\mathbf{r}_o], \tag{6}$$

where p is the dipole moment and $\delta$ the Dirac delta function. Insertion into Eq. (5) yields

$$P = \frac{\omega}{2}\,\mathrm{Im}\{\mathbf{p}^* \cdot \mathbf{E}(\mathbf{r}_o)\}. \tag{7}$$

The electric field in this expression is the field generated by the dipole and evaluated at the dipole's origin. Expressing the field in terms of the Green's function according to Eq. (3), we obtain for the dissipated power

$$P = \frac{\pi\omega^2}{12\varepsilon_o}|\mathbf{p}|^2 \rho_\mathbf{p}(\mathbf{r}_o, \omega), \tag{8}$$

where we used Eq. (2) for $\rho_\mathbf{p}$. Using the known expression for dipole radiation in free space, $P^o = |\mathbf{p}|^2\omega^4/(12\pi\varepsilon_o c^3)$, we can express the LDOS in terms of the normalized power radiation

$$\rho_\mathbf{p}(\mathbf{r}_o, \omega) = \frac{\omega^2}{\pi^2 c^3} P/P^o. \tag{9}$$

The factor $\omega^2/(\pi^2 c^3)$ corresponds to the LDOS in free space and is used in the standard derivation of blackbody radiation.

Comparing Eqs. (1) and (8) we find the interesting result that

$$\frac{P}{\Gamma} = \frac{|\mathbf{p}|^2}{|\langle g|\hat{\mathbf{p}}|e\rangle|^2}\frac{\hbar\omega}{4} \tag{10}$$

The ratio of power dissipation to the transition rate can be expressed in terms of the dipole moments. Failing to distinguish between the transition dipole and classical dipole leads to the erroneous result $P=(\Gamma/4)\hbar\omega$, in which $\Gamma$ represents a photon emission rate. However, $\Gamma$ is not the photon emission rate, but the transition rate between the initially excited electronic state $|e\rangle$ and the electronic ground state $|f\rangle$. The advantage in using the LDOS now becomes clear: it allows for a safe link between quantum and classical formalisms. We now return to the circuit-theory definition of the impedance. The antenna resistance follows from the dissipated power according to Re {Z} =P/I$^2$. Since we have a driving dipole instead of a physical current, it is more useful to define Z in terms of the current density, $\mathbf{j} \sim i\omega\mathbf{p}$, instead of the current, I. Equation (8) thus yields.

$$\mathrm{Re}\{Z\} = \frac{\pi}{12\varepsilon_o}\rho_p(\mathbf{r}_o,\omega). \tag{11}$$

Hence, the LDOS can be associated with the antenna resistance Re {Z}. The units of such a resistance are ohms per area instead of the usual ohms. Notice that Z depends on both the location $r_o$ and the orientation $n_p$ of the receiving or transmitting dipole.

### *Antenna Efficiency, Directivity, and Gain*

The power P in Eq. (8) accounts for the total dissipated power, which is the sum of radiated power $P_{rad}$ and power dissipated into heat and other channels ($P_{loss}$). The antenna radiation efficiency $\varepsilon_{rad}$ is defined as

$$\varepsilon_{rad} = \frac{P_{rad}}{P} = \frac{P_{rad}}{P_{rad} + P_{loss}}. \tag{12}$$

While P is most conveniently determined by calculating the field E at the dipole's position according to Eq. (7), $P_{rad}$ requires the calculation of the energy flux through a surface enclosing both the dipole and the antenna. It is useful to distinguish dissipation in the antenna and the transmitter, which is not accomplished by Eq. (12). We therefore define the intrinsic quantum yield of the emitter as

$$\eta_i = \frac{P^o_{rad}}{P^o_{rad} + P^o_{intrinsic\ loss}}, \tag{13}$$

where the superscripts o designate the absence of the antenna. With this definition of $\eta_i$ we can rewrite Eq. (12) as

$$\varepsilon_{rad} = \frac{P_{rad}/P^o_{rad}}{P_{rad}/P^o_{rad} + P_{antenna\ loss}/P^o_{rad} + [1-\eta_i]/\eta_i}. \tag{14}$$

For an emitter with $\eta_i = 1$ (no intrinsic loss) the antenna can only reduce the efficiency. However, for emitters with low $\eta_i$ we can effectively increase the overall efficiency, which holds promise for the optimization of light emitting devices.

To account for the angular distribution of the radiated power we define the normalized angular power density $p(\theta, \varphi)$, or radiation pattern, as

$$\int_0^{\pi} \int_0^{2\pi} p(\theta, \varphi) \sin\theta d\varphi d\theta = P_{rad}. \tag{15}$$

The directivity D is a measure of an antenna's ability to concentrate radiated power into a certain direction. It corresponds to the angular power density relative to a hypothetical isotropic radiator. Formally,

$$D(\theta, \varphi) = \frac{4\pi}{P_{rad}} p(\theta, \varphi). \tag{16}$$

When the direction $(\theta, \varphi)$ is not explicitly stated, one usually refers to the direction of maximum directivity, i.e.,

$D_{max} = (4\pi / P_{rad}) \text{Max}[p(\theta, \varphi)]$.

Because the fields at a large distance from an antenna are transverse, they can be written in terms of two polarization directions, $\mathbf{n}_\theta$ and $\mathbf{n}_\varphi$. The partial directivities are then defined as

$$D_\theta(\theta, \varphi) = \frac{4\pi}{P_{rad}} p_\theta(\theta, \varphi), \quad D_\varphi(\theta, \varphi) = \frac{4\pi}{P_{rad}} p_\varphi(\theta, \varphi). \tag{17}$$

Here, $p_\theta$ and $p_\varphi$ are the normalized angular powers measured after polarizers aligned in direction $\mathbf{n}_\theta$ and $\mathbf{n}_\varphi$, respectively. Because $\mathbf{n}_\theta \cdot \mathbf{n}_\varphi = 0$, we have

$$D(\theta, \varphi) = D_\theta(\theta, \varphi) + D_\varphi(\theta, \varphi). \tag{18}$$

The influence of an optical antenna on the radiation pattern of a single molecule was recently studied by van Hulst and co-workers, and it was shown that the antenna provides a high level of control for the direction and polarization of the emitted photons.

The gain G of an antenna follows a definition similar to that of the directivity, but instead of normalizing with the radiated power $P_{rad}$ the gain is defined relative to the total power P, i.e.,

$$G = \frac{4\pi}{P} p(\theta, \varphi) = \varepsilon_{rad} D. \tag{19}$$

D and G are usually measured in decibels. Since perfectly isotropic radiators do not exist in reality, it is often more practical to refer to an antenna of a known directional pattern. The relative gain is then defined as the ratio of the power gain in a given direction to the power gain of a reference antenna in the same direction. A dipole antenna is the standard choice as a reference because of its relatively simple radiation pattern. Bouhelier and co-workers recently characterized the relative gain of optical antennas made from metal nanoparticle dimers, using as a reference the dipolelike radiation from single nanoparticles

## *Control of Emission Direction*

One of the most important properties of nanoantennas is controlling the far-field emission from both an emitter and nanostructures. In Fig. 7, a Yagi–Uda plasmonic antenna can beam the emission of a quantum dot positioned near one of the arms of the antenna. Figure 7a shows the SEM images of the nanoantenna composed by five gold nanobars. The distance between each arm in a Yagi–Uda antenna is about 100 nm. A quantum dot is attached to one end of an arm, indicated with a square. The emission direction involves a large solid angle for the quantum dot emitter. Figure 7b shows the radiation intensity distribution at the back focal plane of the objective after an 830 nm long-pass filter. The angular distribution is shown in Fig. 7c, which is consistent well with the simulation result. The constructive interference of the emissions from each arm excited by the emitter results in a narrow directional radiation pattern.

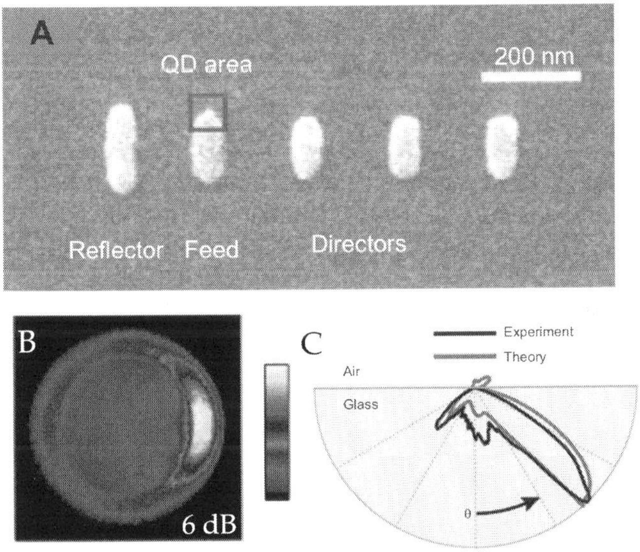


**Figure 7:** (a) SEM image of a five-element Yagi–Uda gold nanoantenna. (b) Radiation pattern from a Yagi–Uda antenna. (c) Measurement and calculation results of the angular radiation distribution for the antenna.

In the Yagi–Uda gold nanoantenna, elements are well separated from each other, which is relatively large for plasmonic coupling.

The emission direction could also be well controlled by stronger coupling. In Fig. 8, the scattering properties can be tuned by two closely spaced metal disks of different materials with appropriate wavelength-dependent permittivities, for example, silver and gold, but the effect is not limited to these materials. The concept is also restricted neither to disk-shaped nanoparticles nor to planar dimers. This bimetallic dimer shows a novel degree of freedom for manipulating the phase or additive phase accumulation by introducing an asymmetric material composition and thus material-dependent SPR. In spite of being as compact as $\sim\lambda^3/100$, this bimetallic nanoantenna shows exotic optical properties, in particular wavelength-dependent directional scattering. Red and blue light will be irradiated in opposite directions, that is, color routing. In the inset of Fig. 8b, corresponding Fourier color images are shown at two specific wavelengths, 450 nm and 700 nm. Interestingly, the direction of the scattering is dependent on the wavelength. For blue (450 nm) and red light (700 nm), the scattering is in opposite directions and the switching point between scattering to the right and to the left appears at around 500 nm. These effects are similar with the incident polarization parallel or perpendicular to the dimer axis. These bimetallic nanoantennas fabricated through cheap colloidal lithography offer a versatile platform for manipulating optical response through wavelength, materials, and geometry, which can be used as ultrasmall ($\sim\lambda^3/100$) photon couplers and sorters, directional emitters, sensors, and metamaterials.

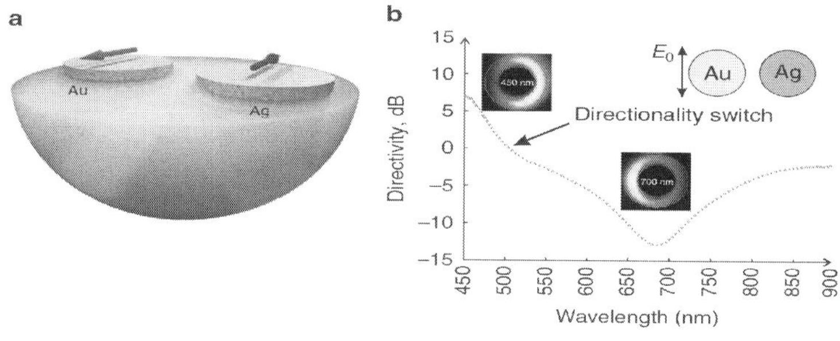


**Figure 8:** (a) An artist's view of a bimetallic nanodisk dimer. (b) Experimental scattering directivities as a function of the wavelength.

## 4.3 PHOTODYNAMIC APPLICATIONS OF NANOANTENNAS

Nanoantennas enable unprecedented control and manipulation of optical fields at nanoscale regions. Due to these properties, they can boost the performance of photo-detection, light emission and photoluminescence. Moreover, they can also enhance the fluorescence and spontaneous emission rates of different emitters, which can lead to new nanophotonic applications. We will present several enhanced photodynamic processes based on nanoantenna configurations and their potential applications.

Emitters, like quantum dots and molecules, exhibit slow fluorescence and spontaneous emission rates. They cannot be used to create high speed optical communication components, such as ultrafast light-emitting diodes (LEDs) and other new optical sources. Recently, it was demonstrated that plasmonic nanopatch antennas are able to increase the inherent slow emission rates of these emitters. To enhance the spontaneous emission rates and to match their operation to high speed optical networks, the plasmonic nanopatch antenna system presented in Figure 9a was used. In this configuration, the nanopatch antenna consists of a nanocube with 75nm size placed over a metallic thin film with thickness 50nm separated by a dielectric spacer layer of thickness 6nm where quantum dots are embedded, as shown in Figures 9a and 9b. The formation of an ensemble of similar single quantum dots placed at the gap of the nanopatch antenna is also presented in Figure 9c.

Modifying the photonic environment of these emitters with the use of plasmonic nanoantennas can lead to a rapid enhancement in their spontaneous emission rate, also known as Purcell factor. The enhancement in the local electric field at the nanopatch's gap can be directly translated to a boosted Purcell factor. The spontaneous emission rate enhancement and the radiative quantum efficiency of this plasmonic nanopatch antenna is calculated with full-wave simulations and is shown in Figures 9d and 9e. Very high spontaneous emission rates are obtained, especially at the corners of the plasmonic nanocube used to form the nanoantenna system. Furthermore, the radiative quantum efficiency remains high (~50%) even with small plasmonic gaps (9nm in this case) due to the exceptionally high field enhancement at the nanogap. The quantum efficiency is calculated as the ratio of radiative over total spontaneous emission rate. It quantifies the fraction of the emitter's energy that is emitted as useful radiation and is not wasted in nonradiative (mainly thermal) processes. Thus, high radiative quantum efficiency is desirable for nanoantennas to be used as ultrafast and broadband optical communication components.

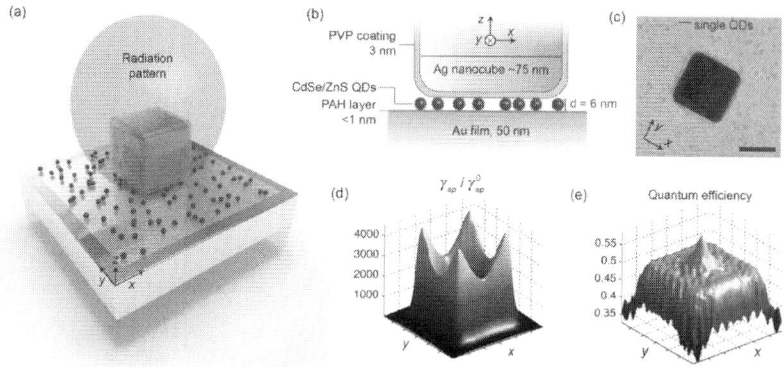


**Figure 9:** Nanopatch antenna with quantum dots. (a) 3D structure. (b) Front view. (c) Random ensemble of single quantum dots combined with the nanopatch antenna. (d) Spontaneous emission rate enhancement distribution. (e) Radiative quantum efficiency distribution.

The dark field scattering and fluorescence spectra of the nanopatch antenna system are demonstrated in Figures 10a and 10b, respectively. In addition, the simulated and experimental scattering spectra of this plasmonic system are measured and shown in Figures 10c and 10d, when the gap of the nanopatch antenna is loaded with a polymer layer or quantum dots, respectively. The fundamental resonant mode exhibits a Lorentzian lineshape in the scattering response when a single nanopatch antenna without quantum dots in the nanogap region is used (see Figure 10c). On the other hand, the scattering spectrum broadens when quantum dots are embedded in the nanogap, as presented in Figure 8d. The broadening in the scattering spectrum is caused due to the inhomogeneous dielectric environment in the nanogap region of the nanoantenna. This nonuniform dielectric environment is resulted due to the random spatial distribution of the embedded quantum dots. Finally, the narrow fluorescence spectrum of this nanoantenna system is also plotted in Figure 10d (red line).

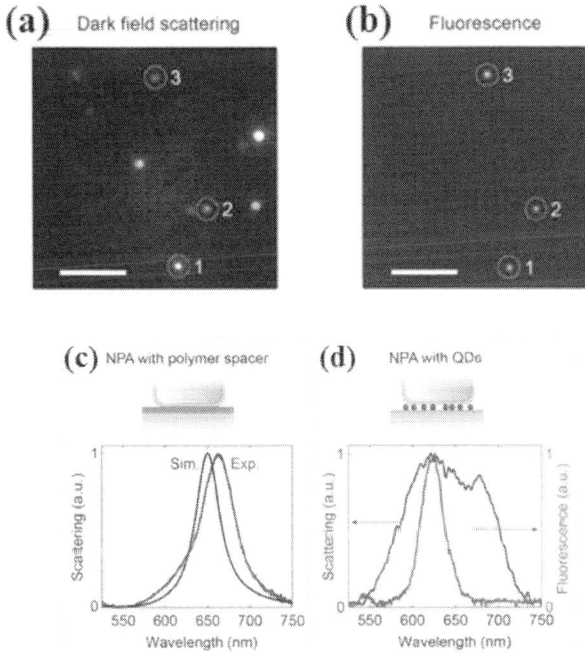


**Figure 10**: Spectral properties of the nanopatch antenna. (a) Scattering and (b) fluorescence images showing individual nanopatch antennas as bright spots with different intensities. (c) The scattering spectrum of the nanopatch antenna with a polymer spacer. (d) The scattering spectrum of the nanopatch antenna with quantum dots embedded in the spacer layer. The fluorescence spectrum is also shown in this plot (red line).

The quantum dot fluorescence intensity is strongly enhanced by the coupling of quantum dots to plasmonic nanopatch antennas. It is presented in Figure 11a as a function of the average incident laser power for three different cases: quantum dots embedded in nanopatch antenna, quantum dots on glass substrate, and quantum dots on gold substrate. The intensity emitted from the quantum dots coupled to a single nanopatch antenna is substantially higher compared to the pure glass and gold substrate cases because strong fields are only induced with the nanoantenna configuration (Figure 11a). These strong fields are the main reason of the fluorescence enhancement in the nanoantenna/quantum dots plasmonic system. Note that quantum dots near the edges of the nanocube geometry experience higher field enhancement due to the shape of the resonant localized plasmon mode formed at the nanogap.

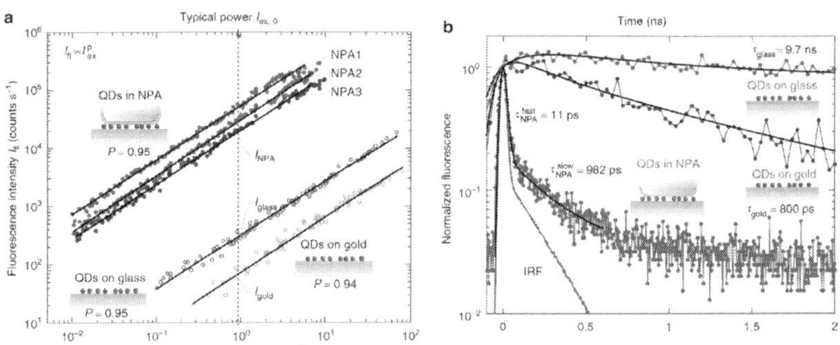


**Figure 11:** (a) Quantum dot fluorescence intensity as a function of the average incident laser power and (b) normalized time resolved fluorescence for three different configurations: quantum dots embedded in nanopatch antenna, quantum dots on gold substrate, and quantum dots on glass substrate.

Finally, the normalized time resolved fluorescence of the aforementioned structures (quantum dots embedded in nanopatch antenna, quantum dots on glass substrate, and quantum dots on gold substrate) is shown in Figure 11b. The enhancement in the spontaneous emission rate can be directly calculated by these time resolved fluorescence measurements. When quantum dots are coupled to the nanopatch antenna, the normalized fluorescence emission is drastically decreased in time compared to just placing quantum dots on glass or gold substrates. The substantial decrease in the fluorescence lifetime can be seen in Figure 11b. It is interesting that this rapid decrease in lifetime is accompanied by a simultaneous increase in the time integrated fluorescence intensity shown in Figure 11a. These are ideal conditions to create ultrafast light sources for new optical communication networks.

Nanoantennas can also be used to enhance other reconfigurable photodynamic applications leading to the creation of efficient and ultracompact photoconductive optical switches, as it is presented in Figure 12. In this case, silver nanorods are used to build the suggested photoconductive nanodipole antenna (Figure 12). They have hemispherical end gap morphology. The nanodipole gap is loaded with amorphous silicon (a-Si) with a large electronic bandgap equal to 1.6eV. This large electronic bandgap is required in order to obtain an ultrafast tunable and reconfigurable response, which is important for optical switching applications. The substrate is not included in this geometry for simplicity.

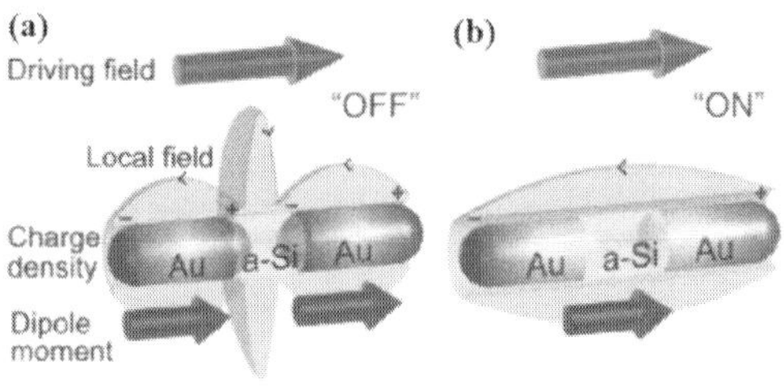


**Figure 11**. Reconfigurable switching operation of photoconductive nanoantenna. Operation at (a) switch "OFF" position and (b) switch "ON" position.

The a-Si material will have different properties when the doping level is increased. The tunable property of the material placed at the nanoantenna's gap will lead to a transition between capacitive and conductive operations. In the unswitched (OFF) case (low doping level of a-Si), the reconfigurable nanodipole antenna supports a half wavelength resonance over each individual arm and not over its entire geometry, as it is shown in Figure 11a. The same nanoantenna in the switched (ON) case (high doping level of a-Si) is shown in Figure 11b. Now, it operates above the free carrier switching threshold limit of amorphous silicon. In this case, the nanoantenna arms are conductively coupled and the nanodipole antenna supports a half wavelength resonance over the entire antenna length, similar to the usual dipole operation.

Nanostructured plasmonic nanoantennas can also be utilized for improved optical sensing and imaging applications, including surface-enhanced fluorescence and Raman scattering. The surface enhanced fluorescence technique is based upon the design of particular engineered surfaces, such as nanoantenna arrays, in the vicinity of different emitters. These surfaces modify and control the local electromagnetic environment surrounding the emitter. This gives rise to an overall improvement in the fluorescence detection efficiency, similar to the nanopatch/quantum dot plasmonic system presented before. Raman scattering can also be boosted by nanoantennas. This process is based on the inelastic scattering of photons, whose energy changes according to the vibration energy of a molecule. Moreover, optical nanoantennas can be used to enhance the efficiency of photovoltaic devices, i.e. for solar energy harvesting applications. They have the potential to improve the absorption of solar radiation and lead to new plasmonic solar cell designs. For example, an array of silicon or plasmonic nanowires can be patterned inside a thin film photovoltaic cell to boost the overall absorption of solar radiation while utilizing less than half of the required semiconductor material.

Another interesting application of optical nanoantennas will be in biological imaging. The schematic illustration of fluorescently labeled antibodies imaged by a nanoantenna probe is shown in Figures 12a and 12b. The fluorescence patterns have different sizes and shapes, which originate from dissimilar antibody aggregates as they lie on the glass substrate. Lastly, optical antenna sensors can be helpful in spectroscopic applications. For instance, the directivity enhanced Raman scattering using nanoantennas was demonstrated.

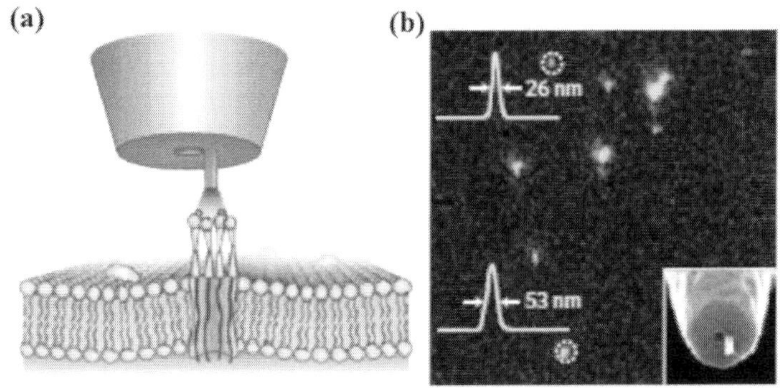

**Figure 12**: Nanoantenna applications in biological imaging. (a) Schematic illustration of imaging with probe based nanoantennas. (b) Florescence imaging with probe based nanoantennas. The geometry of the nanoantenna probe is shown in the inset.

## 4.4 ANTENNAS FOR PHOTOVOLTAICS

An RF or microwave receiver antenna typically generates an alternating current in a transmission line. It is straightforward to rectify this current and produce a direct current output. In fact, the basic self-powered crystal radio receiver is based on this idea. The name rectennas, short for "rectifying antennas," was coined in the 1960s when a project sponsored by the Department of Defense led to the development of a helicopter whose rotor was powered by onboard rectification of outside microwave radiation. The first suggestion to use rectennas for harvesting solar energy came from Bailey in 1972, and Marks subsequently filed several patents describing possible implementations of optical antenna arrays coupled to metal-insulator-metal rectifying diodes.

While conversion efficiencies greater than 50% have been achieved at 10 GHz, even a proof-of-principle demonstration of single-digit efficiencies in the IR or visible regime remains a formidable task. This is primarily because of challenges associated with reliable nanofabrication of antenna arrays coupled to ultrafast diodes capable of optical frequency rectification, and not due to any fundamental physical limitations. It is conceivable that inexpensive and efficient solar rectennas will become a reality in the future by using bottom-up approaches combined with greater control over materials processing, or by employing novel materials such as carbon nanotubes.

The traditional approach to photovoltaics is to use light for generating charge carriers in a semiconductor. The spatial separation of the charge carriers defines a current in an external circuit. For maximum efficiency it is important to absorb most of the incoming radiation, necessitating a minimum material thickness, which forms the primary cost determinant. There are at least three distinct ways in which nanoparticle antennas can interact with a photoactive substrate when placed in close proximity to it, as illustrated in Fig. 13. Plasmonic nanoparticles have large optical cross sections and can efficiently collect and scatter photons into the far field, some of which may become coupled into in-plane waveguide modes in the photoactive material. This leads first to an increased effective optical path length and greater photon absorption probability. Second, the spatially localized high-momentum near-field photons created in the immediate vicinity of an optical antenna can directly excite electron–hole pairs in an indirect gap semiconductor (like silicon) even without phonon assistance, thereby increasing light absorption per unit thickness. Last, there is the possibility of direct charge-carrier injection from the nanoparticle into the semiconductor.

Making use of the first two effects, Stuart and Hall were the first to show increased absorption using metal nanoparticles in a 165 nm thick silicon-on insulator material. More recently demonstrated 16-fold absorption enhancement at $\lambda = 1050$ nm by depositing >100 nm silver particles on thin film silicon-on-insulator cells, while Atwater and co-workers also saw an improvement in the performance of GaAs thin film solar cells by decorating them with silver nanoparticles. The effect of spherical gold nanoparticles on optically thick silicon and found a positive effect most likely due to increased waveguiding. Hägglund investigated thick silicon further by using gold nanodisks and highlighted the problem of increased losses at the plasmon resonance wavelength, which can more than compensate for an increase of absorption in silicon. Improvements in the performance of organic solar cells and InP/InGasAsP quantum-well solar cells have also been reported recently. In all of these examples, the inclusion of nanoparticles in the solar cell structure leads to a reduction in the absorptionmandated thickness requirement for the photoconductive material. A reduced thickness is favorable, especially for organic semiconductors, since it allows a higher fraction of photogenerated carriers to reach the outer world (electrodes) without being lost to undesirable recombination events. A good review of recent developments is given by Catchpole and Polman. Early attempts at incorporating metal nanostructures in Grätzel cells gave mixed results. Hägglund systematically explored this issue by sensitizing a model dye-sensitized solar cell system of flat $TiO_2$ film with nanofabricated gold ellipsoids as shown in Fig. 13. They found evidence for plasmonically enhanced charge generation rate from the dye, which was attributed to a combination of near-field and far-field effects. Adopting a different approach and improving on initial experiments by Kozuka, Tian and Tatsuma have replaced the dye completely by gold nanoparticles and made use of direct photocarrier injection from the nanoparticles into $TiO_2$ to get photon-to-current conversion efficiencies as high as 20%. A scenario combining the two effects of direct charge injection and absorption enhancement of the dye in a nanoparticle-enriched Grätzel cell should be easily realizable.

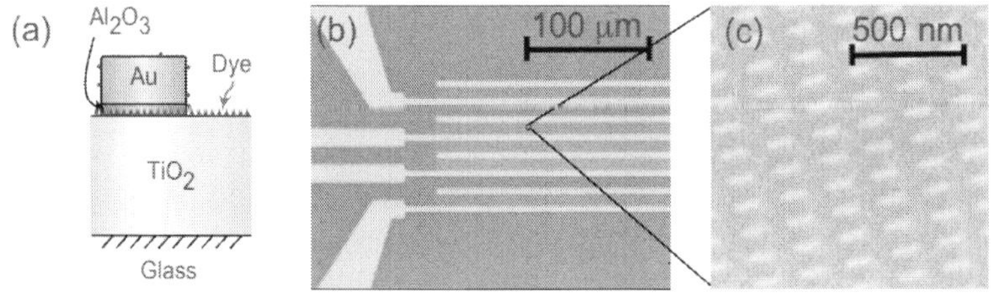

**Figure 13**: A dye-sensitized solar cell decorated with gold nanodisk antennas for improved absorption of light. (a) Schematic of the device. Excitation of the dye molecules adsorbed on $TiO_2$ is amplified near an antenna, leading to enhanced carrier injection from the dyes into the $TiO_2$. (b), (c) Electron micrographs showing electrical contacts and gold nanodisks fabricated on top of thin $TiO_2$.

## 4.5 ANTENNAS FOR LIGHT EMISSION

Because of reciprocity, optical antennas increase not only the efficiency of light absorption but also the efficiency of light emission. They stand to improve inherently low-efficiency lighting schemes like organic LEDs (OLEDs), siliconbased lighting, and solid-state lighting in the yellow and green spectral region.

Typically, an OLED comprises an active organic semiconductor sandwiched between two electrodes. The charges injected from the electrodes recombine as electron–hole pairs (excitons) and then transfer their energy to light-emitting dopants.

Despite the simplicity of the system, OLED technology faces several challenges, and optical antenna structures might provide solutions for some of them. Nominally one quarter of excitons are formed in a singlet (spin-zero) state, and three quarters are in a triplet (spin-one) state. However, in most cases the light-emitting dopants can only convert the energy of singlet states into photons, and the energy associated with triplet states is dissipated as heat.

The triplet states not only reduce the efficiency but are also believed to reduce the longevity of a device. Optical antennas have the potential to quench the triplet state lifetimes, thereby increasing the long-term stability of OLEDs. Another limitation in OLEDs is that some of the energy released in the recombination process becomes trapped in the conducting electrodes as surface plasmons and is eventually dissipated into heat. By patterning the electrodes with arrays of holes, this energy can be released from surface plasmon modes and emitted into free space.

A factor-seven improvement in the electroluminescence quantum efficiency of an OLED device. They attribute this phenomenon to the enhanced transmission of the electroluminescence through the perforated electrode. Perforations in such a metal act as antennas, allowing plasmon modes to couple to propagating radiation.

A similar effect can be achieved by using metallic particles instead of holes and has found success in the field of silicon lighting. The state of silicon light emission lists quantum dots and texture engineering as ways to increase the emission efficiency, but makes no mention of the use of antennas or metallic nanoparticles. Using periodically textured metal films for light extraction from a semiconductor, using an angular emission argument. Who showed a threefold enhancement over the full electroluminescence spectrum of a silicon emitter with silver nanoparticle islands on its surface. Figure 14, which shows a device partially covered with the islands, vividly illustrates the enhancement effect. After one year, they published an improvement to sevenfold enhancement. Their analysis focuses on surface plasmon polaritons in the particles as a coupling mechanism between the incident and local fields.

The particles function as a random array of transmitting optical antennas.

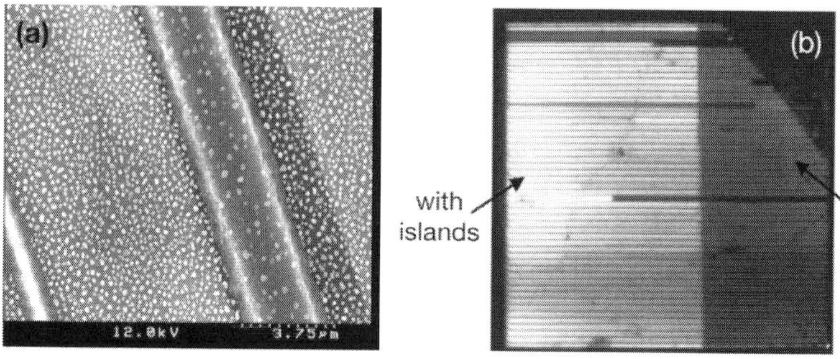


**Figure 14**: (a) Scanning electron microscope image of a silicon-on-insulator LED with silver nanoparticles on its surface. (b) Electroluminescence from an LED partially coated with silver nanoparticles. Brighter areas correspond to stronger electroluminescence intensity

Optical antennas can also be used to enhance the efficiency of solid-state devices at wavelengths where we lack emission-efficient materials. Today's highefficiency solid-state lighting devices use the remarkable properties of LEDs composed of InGaN and AlGaInP. The former has a high emission yield in the blue–green spectral region, and the latter possesses a high emission efficiency in the red. For white light illumination one also requires emission in the green–yellow spectral region, but, unfortunately, InGaN turns out to have a low quantum efficiency in this region.

Near the peak sensitivity of the human eye (wavelength $\sim 550$ nm) the external efficiency is only about 5%. Besides the low quantum efficiency in the green–yellow region, InGaN also possesses an efficiency roll-off at high electron–hole injection densities, limiting the performance of high-power LEDs. One avenue for increasing the emission yield in the green–yellow spectral region is the introduction of phosphors such as $Y_3Al_5O_{12}$ :$Ce^{3+}$ (YAG) that are pumped by the blue emission of InGaN. However, the introduction of phosphors brings other challenges such as saturation and photochemical stability.

Optical antennas in the form of nanoparticles with suitably

engineered plasmonic modes are able to improve the efficiency of low quantum yield emitters and might prove beneficial in solid-state lighting applications. The improvements for a simple spherical nanoparticle antenna and much better efficiency enhancements can be expected for improved antenna geometries. Besides the potential for increasing quantum efficiencies, optical antennas might also enable higher extraction efficiencies by providing a means to extract photons from the source region.

All of these applications involve devices with inherently low efficiency. Antennas can only hurt devices with inherently high efficiency. The limitations of metallic nanoparticles in light-emitting devices. They consider an InGaN/GaN active quantum well in a dielectric medium and model the luminescence with and without a silver nanosphere. Their goal is to optimize the emitter–nanosphere separation and the size of the nanosphere. Nanoparticle antennas may be useful for detection and instrumentation, but that they hold little promise for efficient, large-area emitters.

## REFERENCES


1. Aristotle, De Animalibus, Theodorus Gaza, trans. (Johannes de Colonia and Johannes Manthen, 1476).
2. C. Middlebrook, P. Krenz, B. Lail, and G. Boreman, "Infrared phased-array antenna," Microwave Opt. Technol. Lett. 50, 719–723 (2008).
3. C. Watkins, "ten-," in The American Heritage Dictionary of Indo-European Roots (Houghton Mifflin, 2000), p. 90.
4. D. W. Pohl, "Near-field optics seen as an antenna problem," in Near-field Optics, Principles and Applications, X. Zhu and M. Ohtsu, eds. (World Scientific, 2000), pp. 9–21.
5. E. Prodan, C. Radloff, N. J. Halas, and P. Nordlander, "A hybridization model for the plasmon response of complex nanostructures," Science 320, 419–422 (2003).
6. F. González and G. Boreman, "Comparison of dipole, bowtie, spiral and log-periodic IR antennas," Infrared Phys. Technol. 146, 418–428 (2004)
7. I. Codreanu and G. D. Boreman, "Infrared microstrip dipole antennas— FDTD predictions versus experiment," Microwave Opt. Technol. Lett. 29, 381–383 (2001).
8. M. Schnell, A. García-Etxarri, A. J. Huber, K. Crozier, J. Aizpurua, and R. Hillenbrand, *Nat. Photonics*, **3**(5), 287 (2009).
9. P. Anger, P. Bharadwaj, and L. Novotny, "Enhancement and quenching of single molecule fluorescence," Phys. Rev. Lett. 96, 113002 (2006).
10. R. D. Grober, R. J. Schoelkopf, and D. E. Prober, "Optical antenna: towards a unity efficiency near-field optical probe," Appl. Phys. Lett. 70, 1354–1356 (1997).
11. S. Kühn, U. Hakanson, L. Rogobete, and V. Sandoghdar, "Enhancement of single-molecule fluorescence using a gold nanoparticle as an optical nanoantenna," Phys. Rev. Lett. 97, 017402 (2006).

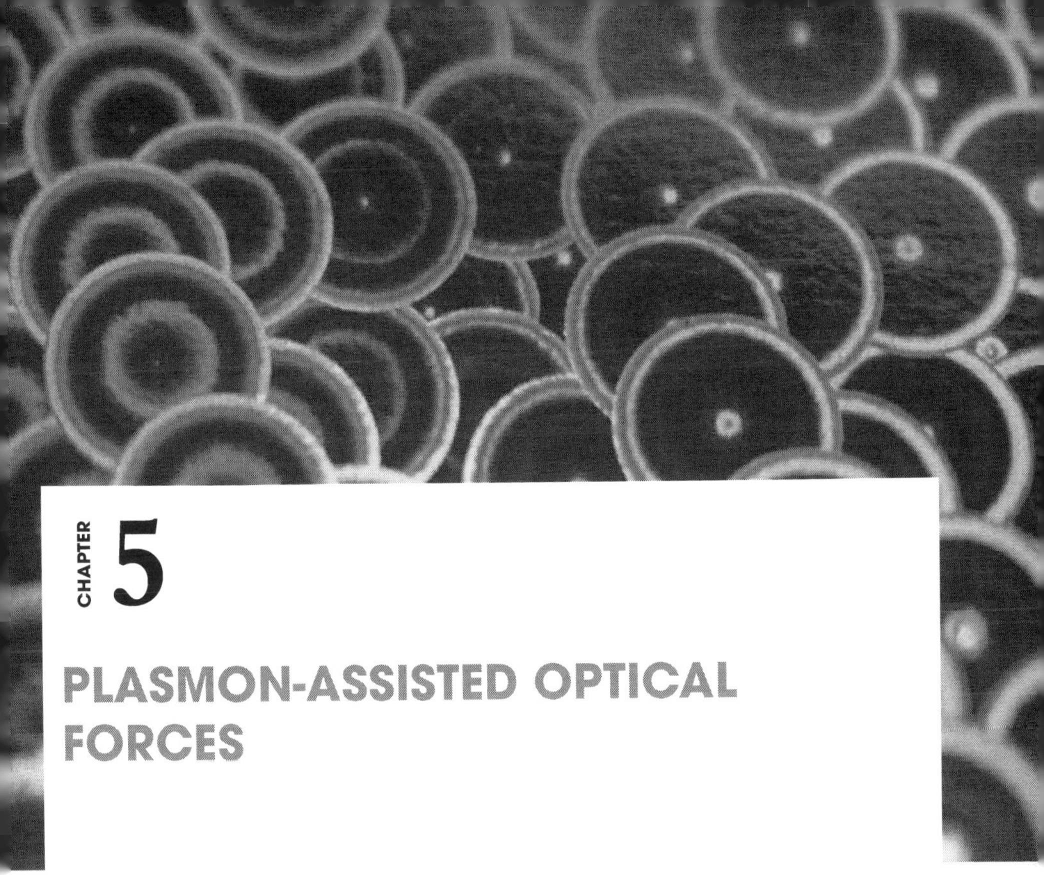

CHAPTER 5

# PLASMON-ASSISTED OPTICAL FORCES

## INTRODUCTION

Optical force has opened up the unique possibility of manipulating objects of micro-/nanoscale sizes by means of light. In general, the forces include a gradient type, which originates from the electric field gradient of light, and a dissipative type, also known as the radiation pressure, a result of photons impinging on and absorbed by the objects. A typical application of optical forces is the so-called optical tweezer that was discovered by Ashkin in 1971 and have nowadays become a powerful tool in optical trapping and manipulation, for example, sorting biological cells, measuring biophysical properties of single biomacromolecules, and manipulating dielectric and metallic nanoparticles. In nanostructures that support surface plasmons, an enhanced electromagnetic (EM) near field is induced under resonant excitation, leading to a series of interesting applications in surface-enhanced spectroscopy and sensing. The induced EM fields are closely bound to the surface and attenuate dramatically with increasing distance from the surface, thus forming a sharp gradient that gives rise to an attractive optical force proportional to the gradient. For propagating surface plasmons, the scattering force is also an interesting component to study—it pushes the trapped particle to move along the propagation direction, for example, on the surface of metal thin films and nanostripes.

Plasmon-assisted optical forces can be discussed in two slightly different categories: with and without a focused laser beam that provides additional gradient force from the laser itself. The former mainly refers to laser trapping of metal nanoparticles, where surface plasmons take effects in such a way that the behavior of the trapped particles differs substantially compared to dielectric ones and the interaction between the trapped metal nanoparticles involves the optical component resulting from the near-field coupling. The latter deals with optical trapping and manipulation using metal nanostructures fabricated on a substrate. In this case, an unfocused or slightly focused laser beam is used only to excite surface plasmons in the structures but not to provide any additional trapping effect.

There have been a number of theoretical and experimental work in the field of surface-plasmon-assisted forces, for example, optical forces of spherical metal nanoparticles in an optical trap of laser tweezers, optical trapping of dielectric/metal nanoparticles using metal nanostructures or metal thin films, and plasmon-driven nanomotors.

## 5.1 THEORETICAL CALCULATIONS ON OPTICAL FORCES IN NEAR-FIELD -COUPLED NANOPARTICLES

Great efforts have been made in the theoretical calculation of optical forces of metal nanostructures, for example, spherical and bipyramidal metal nanoparticle dimers and aggregates of metal nanoparticles, single and dimers of nanodisks, and metal thin films. Typically, the EM field is first calculated using electrodynamic methods such as generalized Mie theory (GMT), finite-difference time-domain (FDTD), or discrete dipole approximation (DDA). The force can then be obtained using Maxwell's stress tensor formalism or the Lorentz force method. To define the stability of an optical trap, the optical potential can be calculated by integrating the optical force from the point of interest to infinitive and is usually given in the unit of thermal energy at room temperature $k_B T$, where $k_B$ is the Boltzmann constant and T is the room temperature.

In 2002, the theoretical calculations of optical potential and forces between spherical metal nanoparticles has been showed. In Figure 1a, the spatial variation of optical potential U/kBT around a chain of three nanoparticles in water is demonstrated. The radius of Ag nanoparticles is 25 nm, and the gap distances are both 1 nm. It was excited at resonance by a 760 nm laser at 10 mW/μm$^2$ power density. It is seen that the strength of the optical potential is the strongest at the nanogap where the EM field is the strongest. Accordingly, it can be anticipated that a single molecule that resides within or passes through the trapping volume would be instantaneous trapped at the gaps. The optical force at the gap of a dimer of Ag nanoparticles (schemed in the inset: radius 45 nm and gap distance 1 nm) under resonance illumination (wavelength 550 nm) is shown in Figure 1b. FC and FD represent the gradient and dissipative forces, respectively. The direction of the forces can also be found, that is, the gradient force always points to the gap center (positive to the left and negative to the right), while the scattering force is always along the incident k-vector (positive values). On the other hand, the gradient force is more than 1 order of magnitude stronger than the scattering force. These results provide theoretical support for single-molecule surface-enhanced Raman scattering (SERS) in the sense that even though the exact location of a single molecule is not certain, the molecule could be spontaneously attracted to the so-called hot spots where the EM field is the strongest due to plasmon-induced optical forces.

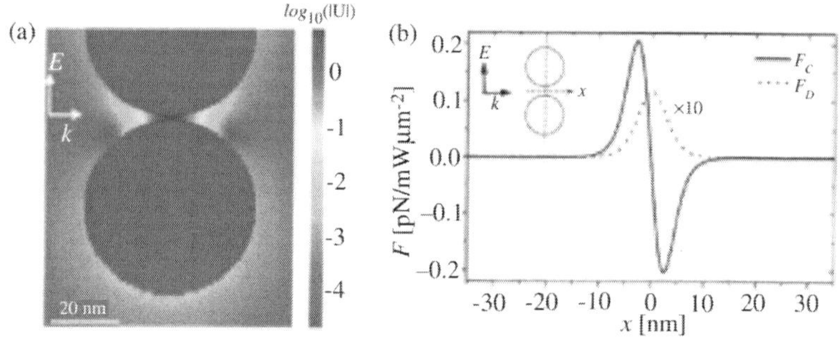


**Figure 1**. (a) Spatial optical potential distribution around a trimer of silver nanoparticles with 25 nm radius and 1 nm gap distance in water excited by the resonant wavelength (760 nm) at a power density of 10 mW/μm$^2$. (b) Optical force distribution at the gap of a dimer of Ag nanoparticles (radius 45 nm and gap distance 1 nm) along the incident wave vector (see inset) under 550 nm excitation.

For near-field-coupled metal nanoparticles, the optical force is enhanced because of the enhanced EM field gradient. This naturally leads to another interesting phenomenon: the spontaneous dimerization/aggregation of metal nanoparticles under external illumination. Figure 2 shows the polar plot of the total optical force on the surface of the lower particle of the same dimer as in Figure 1b. It is obvious that the force field is highly concentrated to the gap center, meaning that the particle itself experiences a net attractive force that would bring the two particles even closer. For comparison, the magnitude of the net force after surface integration is 62 pN under 1 mW/μm$^2$ toward the other particle in the dimer, while the radiation pressure force along the k-vector is only 0.3 pN. An attractive force is expected between spherical nanoparticles of the same size excited by a parallel polarized laser. However, a repulsive force is also theoretically predicted under certain excitation conditions for a dimer of two identical particles or heterodimers. Miljkovic calculated the optical force of a dimer of 5 nm Ag nanoparticles under both parallel and perpendicular polarizations. Although the two methods.

(Maxwell's stress tensor and Mie theory [MST-Mie] shown by the thick solid lines and Lorentz force–coupled dipole approximation

[LF-CDA] shown by the thin solid lines) produce slightly different results, the trends are the same. Under parallel polarization, an attractive force (negative values) is induced between the particles, in consistence with previously reported results. Interestingly, a repulsive force appears if the laser polarization is perpendicular to the dimer, although it is about 1 order of magnitude weaker than the attractive force in the former case. The repulsive force is easy to understand from a simple picture of dipole–dipole interaction. Neglecting retardation, two identical dipoles are in phase with each other under parallel polarization, leading to an attractive force. With perpendicular polarization, the two dipoles are out of phase, meaning that the sides with the same type of charges face each other, resulting in a repulsive force. A repulsive force can also be produced under parallel polarization for heterodimers, that is, a dimer of nonidentical spheres. In this case, the attractive and repulsive component of the optical force can even cancel out at certain separations and excitation wavelength, indicating that an equilibrium position exists between the particles. Mie theory considers a large number of multipoles in the particles and is more accurate than the CDA method, which is dipole approximation based.

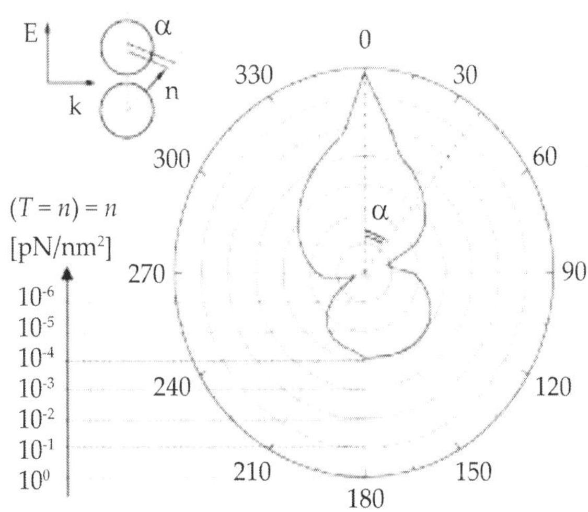


**Figure 2.** Polar plot of the optical force on the surface of the lower particle in a dimer under irradiation of 1 mW/μm² at 514.5 nm.

The theoretical calculations assume that the structure of interest is under irradiation of a plane wave. From a practical point of view, a Gaussian beam is widely used in an optical tweezer system and thus would be more interesting to study theoretically. Li calculated the optical potential experienced by a metal nanoparticle in an optical trap of a focused laser beam in the presence of an immobilized one. The Gaussian beam was expanded into a sum of vector spherical harmonics (VSHs) with different orders by using the Davis formalism. Figure 3 shows the optical potential of a free silver nanoparticle of radius R as a function of the x position to the focus of a Gaussian beam, as illustrated in the inset, where F and I represent the free and immobilized particles, respectively, and D is the distance between the laser focus (the origin of the coordinate) and particle I. The incident laser wavelength is 514 nm for the green curves and 830 nm for the red curves. The results for particle radii 20 nm, 50 nm, and 100 nm are shown in Figures. 3a, b, and c, respectively. Taking the case of particle I being placed at the center of the beam (D = 0 nm, represented by the thin curves) as an example, it can be seen that for all the particle sizes, the optical potential of particle F drops dramatically when F approaches I, indicating that two particles can be easily dimerized in an optical trap of a focused laser beam. The left insets in the figure show the van der Waals potential between the two nanoparticles as a function of separation, which is comparable to the optical potential only for large particles at extremely short distances (e.g., 1–2 nm). The van der Waals force is an important factor that one has to consider while studying the interparticle forces in an optical trap.

In general, the scattering force and absorption force point at the direction of the incident wave vector, and have been considered detrimental for an optical trap, particularly in 3D trapping. Indeed, the absorption annihilates the incident photons and the resultant momentum is, of course, along the incident direction.

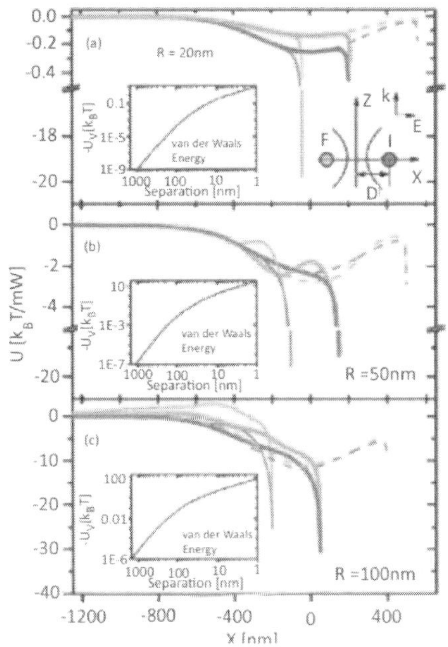


**Figure 3**. The optical potential experienced by a free silver particle F close to an identical immobilized particle I as a function of the x position.

However, the scattering component is more complicated. The magnitude and direction of the scattering force depend on the EM modes that are excited in a particle, so the scattering force could be used for trapping instead. Tractor beams that drag the particle moving against the incident wave vector have been theoretically demonstrated using shaped laser beams such as the Bessel beam. Even for a Gaussian beam, it is possible to produce an attractive scattering force on a metal nanoparticle. For a gold nanoparticle (GNP) placed off-axis in the focal plane, if the wavelength of the incident laser is close to the quadrupole resonance, the interference between the dipole mode and the quadrupole mode results in stronger outward scattering than the inward scattering along the laser polarization. As a result, an attractive scattering force is exerted on the particle. Such an effect can in principle be used for size selective 2D optical trapping in the focal plane. We shows the calculated results for trapping wavelengths of 410 nm and 830 nm. It is obvious that the 410 nm laser produces an optical potential only for particles of radii around 60 nm and 100 nm and is not able to trap particles of other radii. For comparison, the 830 nm laser can trap particles of all radii.

## 5.2 EXPERIMENTAL DEMONSTRATIONS OF PLASMON-ASSISTED OPTICAL FIBERS

Optical fiber is a strand of transparent dielectric waveguide, which can guide and transmit optical signals with extremely low loss. It has a lot of merits in being used as a platform for the excitation of surface plasmon polariton (SPP), which is the collective oscillation of electrons at a dielectric-metal interface, and for its applications from the perspectives of compactness, flexibility, and cost-effectiveness.

### 5.2.1 Optical Forces on Metal Nanoparticles trapped by a focused Laser Beam

Optical tweezers have been widely used to trap dielectric and metal nanoparticles. The distinct difference between dielectric and metal nanoparticles lies in the polarizability. Due to the larger polarizability, metal nanoparticles experience much higher optical forces, including the gradient, scattering, and absorption forces. On the other hand, the stable trapping of metal nanoparticles provides a platform for the study of physical properties of single metal nanoparticles and interactions between multinanoparticles. It also opens up the possibility of surface-enhanced spectroscopic sensing due to the near-field coupling effect between the trapped metal nanoparticles.

#### *Elongated Nanoparticles*

Metal nanorods exhibit a polarizability difference along the long and short axes simply due to the different spatial confinement of the oscillation of free electrons. In an optical trap of a focused near-infrared (NIR) beam, the long axis of a metal nanorod turns out to self-align parallel to the laser polarization to minimize the potential energy of the induced dipole, given by $U = -\langle \mathbf{p} \cdot \mathbf{E} \rangle = -1/2 \sum \mathrm{Re}\{\alpha_{ii}\} E_i^2, i = x, y, z$, where p is the induced dipole moment, E is the incident electric field, and $\alpha_{ii}$ is the polarizability . Furthermore, if the polarization of the trapping laser is rotated by an angle of β, the nanorod experiences a torque that equals

$|\langle \boldsymbol{\tau} \rangle| = |\langle \mathbf{p} \times \mathbf{E} \rangle| = 1/4 E_0^2 (\alpha_{//} - \alpha_{\perp}) \sin 2\beta$ where $\alpha_{//}$ and $\alpha_{\perp}$ are the polarizability components parallel and perpendicular to the long axis of the nanorod, respectively. This torque drives the nanorod to realign parallel to the laser polarization and rotate if the linear polarization is further rotated.

An example of a trapped Ag nanorod is shown in Figure 4a. One sees a peak at ~500 nm in the dark-field (DF) scattering spectrum when the white-light polarization is parallel to the laser polarization and a peak at ~460 nm when the polarization is perpendicular. This indicates that the nanorod indeed aligns with its long axis parallel to the laser polarization. Figure 4 b shows the polarizability plot for a silver nanorod of 60 nm diameter and 80 nm length in water as a function of the wavelength calculated using the modified long-wavelength approximation (MLWA). It is clearly seen that the polarizability of the nanorod is much higher at 830 nm, when the polarization of the laser is parallel rather than perpendicular to the long axis of the nanorod. Hence, a stronger dipole, and consequentially a deeper potential well, is induced if the long axis is parallel to the electric field of an 830 nm laser. Figure 4c demonstrates the rotation of the nanorod by rotating the linear polarization of the trapping laser. The polarization of white-light illumination is fixed while the polarization of the trapping laser is rotated up to 90°. It is seen from both the DF scattering images and spectra that the scattering intensity decreases when the laser polarization is rotated from parallel to perpendicular to the white light polarization, implying that the nanorod is indeed rotated. It is worth noting that a dimer of spherical metal nanoparticles in principle behaves as a nanorod, with its dimer axis as the "long axis," and thus also aligns parallel to the laser polarization and can be rotated by rotating the laser polarization.

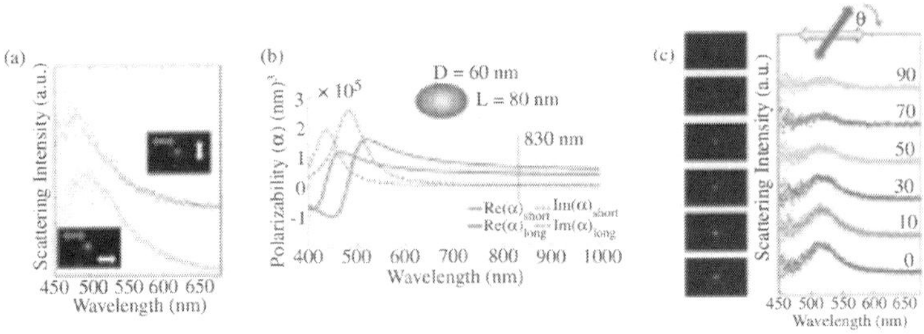


**Figure 4**. Alignment and rotation of a trapped Ag nanorod. Red and white arrows in (a) and (c) represent the polarization of the NIR trapping laser and white light, respectively. (a) Dark-field (DF) scattering spectra and images of a trapped nanorod. (b) Calculated polarizability of a nanorod (D = 60 nm and L = 80 nm). (c) DF images and spectra of a trapped nanorod at different angles between the laser polarization and the fixed white-light polarization.

It has been reported that semiconductor nanowires of several microns in length can be trapped and manipulated to build nanostructures.

Shelton reported the optical trapping of glass nanowires in water and investigated the alignment and rotation behavior. They showed that a glass nanowire tends to align along the propagation of the trapping laser (514 nm) or parallel to the electric field if the self-alignment along the propagation is hindered. Silver nanowires in an optical trap also align themselves but, interestingly, perpendicular to the laser polarization. In Figure 5a, a silver nanowire of length ~5 μm is trapped at one of its tips (highlighted by the circle). The arrows on the right and left represent the polarization of the trapping laser and illuminating white light, respectively. Experimental observations show that, on the one hand, the preferential trapping spot is either of the ends of the nanowire, and on the other hand, more interestingly, the nanowire aligns perpendicular to the laser polarization. DDA simulations of the near fields generated by the wires under irradiation of an 830 nm plane wave were performed to understand the observations. The simulation results are shown in the right panel of Figure 5b. For comparison, the near fields of a silver nanorod (D = 40 nm and L = 80 nm) are also given in the left panel. It is seen that, for a nanorod, the near fields are much stronger at parallel configuration than that at perpendicular configuration. However, the nanowire case is completely different. Much stronger near fields are induced when the laser polarization is perpendicular to the nanowire rather than parallel. This means that the nanowire tends to align perpendicularly to the laser polarization, which was observed experimentally. Another important feature is that the strongest field is always at the tips of the nanowire. This in principle explains the end-trapping of nanowires. Again, the rotation of the laser polarization produces an optical torque that drives the nanowire to rotate, as shown in Figure 5.a-I to 5a-III. The fixed nanowire to the left can be regarded as a reference.

By continuously rotating the linear polarization of the trapping laser, the trapped particles spin accordingly. This is of particular interest from an application point of view. In a solution, due to the viscous hindrance of the surrounding medium, the axis of an elongated nanoparticle (a nanowire, a nanorod, or a dimer of spherical nanoparticles) is expected to lag behind the laser polarization. Such a phase delay strongly depends on the surface chemical properties of the particles, the viscosity of the surrounding medium, and the strength of the surface-plasmon-induced optical torque. Although the phase delay is difficult to observe on nanorods and dimers of spherical nanoparticles, perhaps due to the weaker drag force with respect to the surface-plasmon-induced torque, the lag is obvious in case of nanowires. For a nanowire of ~5 μm length and ~100 nm diameter, a phase delay of ~30° has been found at a rotation frequency of 0.8 Hz.

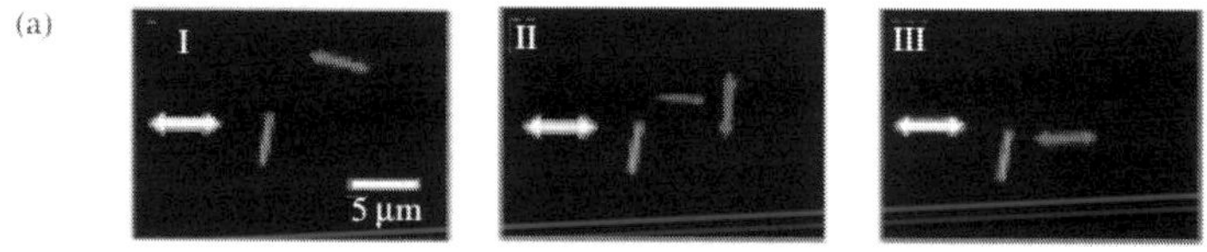

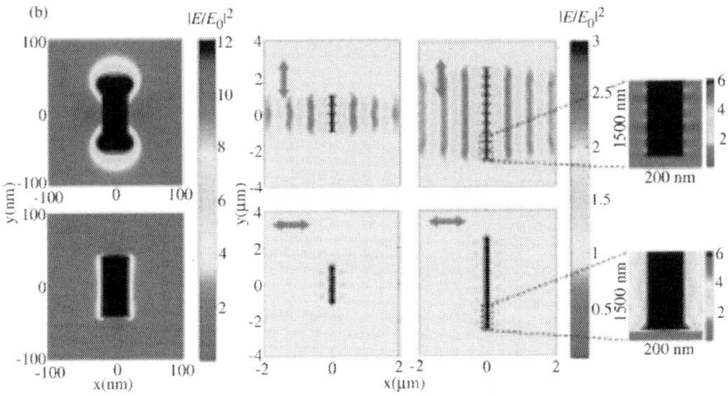


**Figure 5**. Alignment and rotation of a trapped Ag nanowire of ~100 nm in diameter and ~5 μm in length.

By applying a circular or elliptical polarization, the nanowires can even spin spontaneously. This is due to the direct transfer of the angular momentum from the incident photons to the nanowire. The spinning frequency is then proportional to the laser power at a specific polarization status. The direction of rotation can be easily changed by changing from right- to left-circularly-polarized light. One should point out that any anisotropic nanoparticles, such as nanorods and dimers of spherical nanoparticles, are also expected to spin spontaneously under irradiation of a circularly or elliptically polarized laser.

It is worth noting here that asymmetrical metal nanoparticles can also be rotated and spontaneously spin under illumination of a linearly polarized plane wave or a focused laser beam. Liu fabricated a gold nanomotor (100 nm radius, 380 nm line width, and 30 nm thickness) embedded by a silica microdisk that is 4000 times large in volume, which can spin when shined by a linearly polarized NIR laser. The direction of spinning can be changed simply by changing the laser wavelength. This is supported by their calculation of optical torque over 600–2400 nm wavelength range, which shows positive torques below ~1200 nm but negative values above.

## *Interaction between two metal nanoparticles in an optical trap*

The understanding of the interactions between individual nanoparticles is crucial in the control of colloidal stability and in the rational guide of applications such as building blocks of self-assembled nanoparticles. In an optical trap, two metal nanoparticles experience several forces that determine the response of the dimer. We will show that the surface-plasmon-induced optical force plays an important role in probing the physical properties of metal nanoparticles in a colloid at the single-particle level.

In a metal colloid, the particles are typically negatively charged due to the adsorption of anions, thus producing an electrostatic repulsion between adjacent particles. This repulsive force depends on the number of surface charges and can be adjusted by changing the ionic strength of the solution. On the other hand, the van der Waals attractive force is determined by the material-specific Hamaker constant and particles' sizes and separation. The two forces, the Coulomb repulsion and van der Waals attraction, are the main components of the well-known Derjaguin–Landau–Verwey– Overbeek (DLVO) forces, which are used to interpret the colloidal stability. The total potential is then the sum of DLVO and interparticle optical potentials.

Two GNPs of diameter 80 nm were trapped and dimerized in an optical trap of a NIR laser (830 nm). We show the DF scattering images and spectra, respectively. The final concentration of salt (NaCl) in the colloid is adjusted to be 20 mM. The white and red double arrows represent the polarization of white light for DF illumination and the trapping laser, respectively.

They show that the scattering color does not change with the change of white-light polarization, indicating the spherical shape of the particle. However, the DF scattering of a dimer of particles shows clear polarization dependence.

If the laser polarization is parallel to the white-light polarization, the surface plasmon resonance peaks at ~620 nm due to the near field coupling and the dimer is seen as reddish in DF imaging If the polarizations are perpendicular, the resonance peak is at ~560 nm, which is close to the resonance of a single particle.

We show the calculated potential curves of the dimer with (lower curve) and without (upper curve) optical potential. The dominant features in the curves are a deep primary minimum at near-zero separation and a potential barrier that prevents irreversible aggregation. Another important feature is that there is a shallow secondary minimum at ~15 nm separation without optical potential. For some colloids and ionic strengths, this secondary minimum may result in spontaneous aggregation or flocculation, but here the depth is ~1 kBT and not deep enough to induce stable dimerization against the thermal Brownian motion. Consequently, the DLVO potential curve indicates that such a colloid should be stable against aggregation up to a salt concentration of at least 20 mM. However, as illustrated by the lower curve, the secondary minimum deepens dramatically if the laser induced attractive optical potential is added. For a laser power of only 2 mW/$\mu m^2$ at $\lambda = 830$ nm, a potential well depth of ~5 kBT at ~9 nm separation was obtained. This leads to stable dimerization and a significant red shift in DF scattering due to strong plasmon hybridization.

For a colloid, adding salt leads to an increased screening of the surface charges through a build-up of the density of counterions on the particle surfaces and within their immediate surroundings, the so-called electrical double layer. As a result, the electrostatic repulsion should decrease with increasing salt concentration and the position of the secondary potential minimum should move toward smaller separations, leading to an increased near-field coupling and spectral red shift for a dimer of GNPs. Such a concentration dependent red shift is clearly shown.

The measured peak positions (circles) of the longitudinal mode of the dimers are plotted for different NaCl concentrations of 0, 5, 10, 15, and 20 mM. On the other hand, the perpendicular plasmon mode remains close to the single-particle resonance position.

From a theoretical point of view, the total potential can be approximated as $U_{tot} = U_c + U_{vdW} + U_{opt}$, where the first two terms, the Coulomb repulsion potential and van der Waals attraction potential, constitute the DLVO potential and the third is the optical potential. The optical potential $U_{opt}$ is the sum of two terms $U_{inter}$ and $U_{trap}$, that is, the potentials due to interparticle coupling and particle confinement within the laser focus, respectively. The latter term turns out to be much smaller than the interparticle potential and can be safely neglected. The former item, $U_{inter}$, can be calculated using the Maxwell's stress tensor formalism combined with Mie theory as described in the theoretical calculations section.

The theoretical scattering spectra can be obtained from the numerical integration of the product of normalized probability distribution and the calculated spectra at a certain separation. The peak positions can then be found and are shown by the crosses together with a linear fit (black line), which finds an excellent agreement with the experimental results. The minimum mostprobable separation of the two particles, as well as the depth of the secondary minimum potential well, can also be obtained an. Although there is a considerable variation in the red shift between different particle pairs measured under identical external conditions, probably due to the difference in surface charge densities of particles, the methodology described herein provides a valuable tool for characterizing nanoparticle heterogeneities that are not accessible through ensemble averaged measurements.

### *Applications in SERS sensing*

Optical forces induce spontaneously aggregation of metal nanoparticles, which in turn leads to strong near-field coupling and produces high field enhancement in the hot spots. Such aggregates are excellent candidates for SERS applications.

A dimer of two metal nanoparticles created optically in a focused laser beam has shown high enough field enhancement for efficient SERS detection. The example shows the dimerization of two silver nanoparticles using optical tweezers to create hot spots for SERS. In these experiments, a NIR laser of 830 nm is used for optical trapping and a separate laser at 514.5 nm is used for Raman excitation. Thiophenol molecules are chemically bound to the surface of the silver nanoparticles as Raman probe. A single silver nanoparticle immobilized on a glass slide or trapped by the focused laser can be identified by DF imaging as a bluish spot, from which no SERS signal is detected (insets in the top and middle panels). However, if a dimer is created by moving a trapped particle to an immobilized one, a significant SERS signal arises due to the strong near-field coupling effect (see the inset in the bottom panel).

We show the calculated optical potential experienced by a silver nanoparticle that is free to move in a Gaussian laser focus at 830 nm wavelength and 50 mW laser power. The white circle represents an immobilized particle that is 250 nm away from the laser focus along the x axis.

The result is similar, where a deep potential minimum is induced when the trapped particle approaches the immobilized one, giving rise to spontaneous optical dimerization and a SERS hot spot in the optical trap, which is responsible for the appearance of a SERS signal after dimerization.

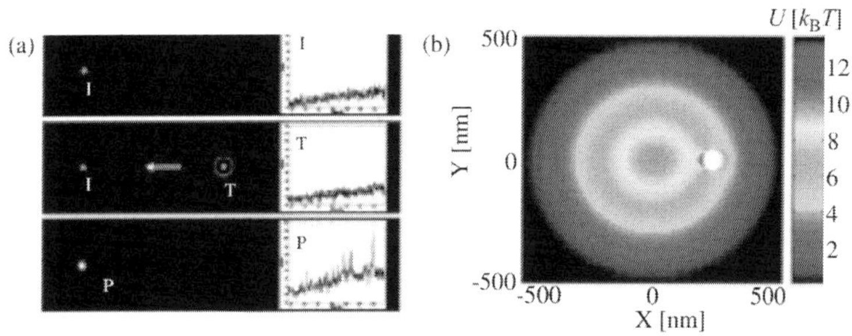


**Figure 6.** (a) The dimerization of silver nanoparticles, DF images, and corresponding Raman spectra. (b) Simulated optical potential under 830 nm laser irradiation experienced by a trapped Ag nanoparticle with an immobilized one that is 250 nm away from the laser focus.

For larger aggregates, stronger enhancement is expected. The work by Tanaka showed that multiple silver nanoparticles simultaneously trapped by a NIR laser beam also tend to align parallel to the linear polarization and that, more importantly, the aggregates exhibited pronounced SERS signal at a very low concentration of probe molecules, $10^{-14}$ M pseudoisocyanine (PIC), excited by a separate laser beam, whereas no SERS appeared if the trapping laser was not focused in the solution.

Optical-force-induced aggregation in laser tweezers has prominent potential for lab-on-a-chip-based chemo-/biosensing applications. In a microfluidic channel, the metal nanoparticles incubated by the analyte molecules can be trapped in the flow and form aggregates, which are ideal substrates for SERS sensing. Figure 7a shows DF images of the growing aggregate created by optical tweezers using NIR laser trapping at 830 nm. The corresponding SERS spectra, excited by a 514.5 nm laser line, are shown in Figure 7. With a growth in size, the aggregate shows increased SERS enhancement. Using microfluidics, the analyte flow is interchangeable, enabling consecutive detection of a series of species simply by changing the analyte solutions and keeping the Ag colloid flow. With a Y-shaped microfluidic channel, where metal colloid and analyte solution can flow in parallel, a SERS signal can also be detected at the interface where the nanoparticles and the analyte molecules are mixed through diffusion.

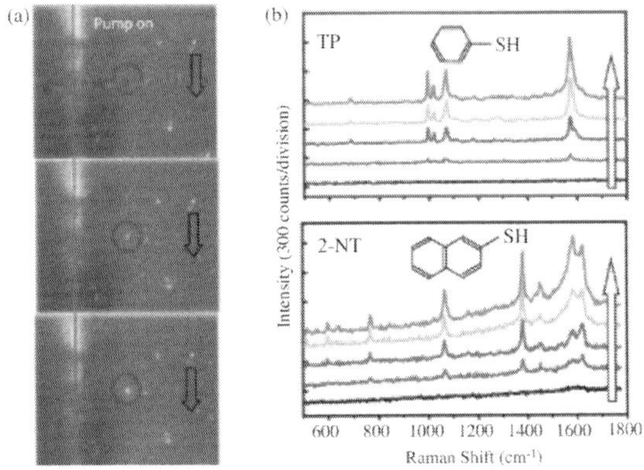


**Figure 7.** (a) DF scattering images of silver nanoparticles trapped in a microfluidic channel. The trapping wavelength was 830 nm. (b) Corresponding SERS spectra from the optically induced nanoparticle aggregates.

### 5.2.2 Optical Forces in Lithographically Fabricated Plasmonic Nanostructures

By using metal nanostructures fabricated on a substrate, Plasmon assisted optical forces can be used to manipulate objects in a more controllable way. An advantage of such experiments is that the laser power density can be much lower than that used in optical tweezers and thus would cause a much less heating effect resulting from absorption and less damage in biological applications.

#### *Gold nanopads and nanoholes*

For gold nanopads fabricated on a glass substrate, if excited by total internal reflection illumination, the optical potential experienced by a dielectric nanoparticle exhibits a potential well centered at the forward edges of the nanopads. Figure 8a shows the map of optical potential of a 200 nm polystyrene (PS) bead located 20 nm above a 450 nm × 450 nm gold pad of 40 nm thickness under 785 nm laser irradiation of power density 5 × 107 W/m$^2$. The asymmetrical distribution of the potential mainly arises from the strong forward scattering due to the illumination asymmetry. Nevertheless, the depth of the potential well is stable enough to trap the PS bead, as demonstrated in Figure 7b, where arrays of patterned gold nanopads are used.

Besides metal nanodisks, nanoholes in a metal film can also trap dielectric beads under normal illumination at resonance. It has been shown that PS beads of 50 nm and 100 nm can be trapped in cylindrical apertures of a diameter of a few hundred nanometers in a 100 nm thick gold film using a 1064 nm laser. The laser power is less than 2 mW, significantly lower than that used in conventional optical trapping experiments with laser tweezers. Figure 8c shows the scheme of the experiments. As shown in Figure 8d, the transmitted intensity of the laser fluctuates during the trapping process. If a PS nanoparticle is trapped at the edge of an aperture, the transmission increases. Although the particle tends to escape due to the thermal motion, the displacement of the bead would result in a restoring force that pushes it back, which the authors called self-induced back-action (SIBA).

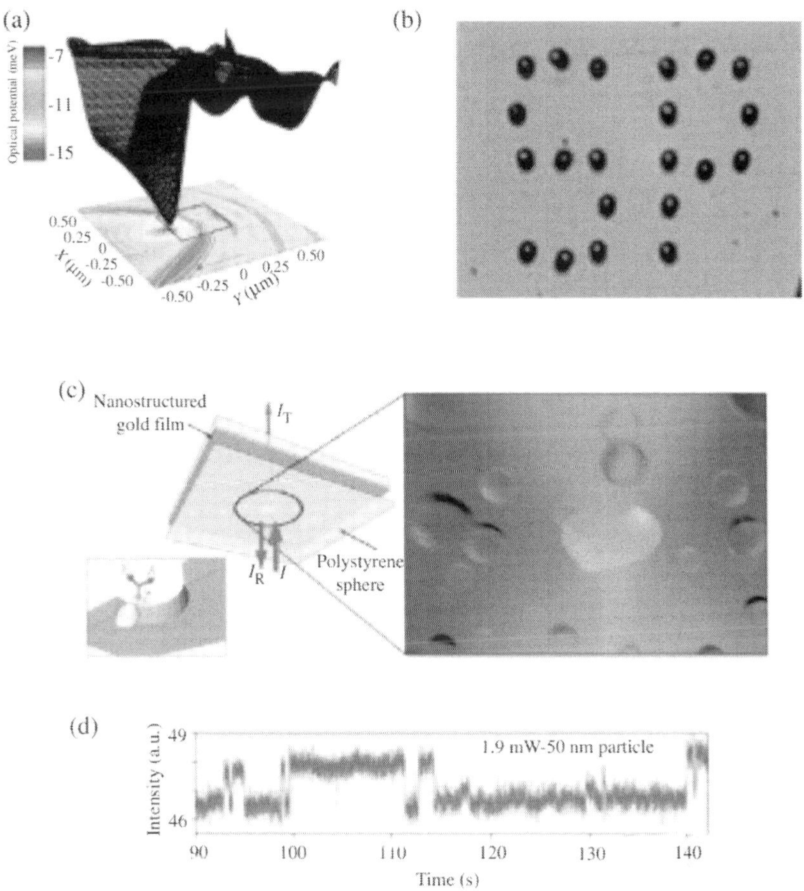


**Figure 8**. (a) Theoretical optical potential experienced by a 200 nm PS bead placed 20 nm above a gold nanopad. (b) PS beads trapped on a patterned array of gold nanopads. (c) Scheme of optical trapping using nanoholes in a gold film. (d) Intensity fluctuation of the laser focused on a nanohole due to the optical trapping of a PS bead.

## *Dimers of Nanodisks*

The coupling between metal nanostructures produces extra field enhancement in the gaps, which consequently enables more precise optical trapping and positioning of nanoscale objects. Compared to a conventional optical tweezer setup, the trapping efficiency can be greatly improved and the Brownian motion of the trapped object can be suppressed by almost 1 order of magnitude. Figure 9a shows the recorded positions of a trapped single PS bead with and without the presence of a gold nanodimer (time step: 5 ms). In conventional 3D optical tweezers without gold nanodisks (top-left scheme), the movement of the PS bead is confined within the trap created by the laser itself, which is a circle of ~1 μm diameter (green solid circles). However, if the trapped bead is close to a dimer of gold nanodisks (top-right scheme), it experiences a strong local electric field and is attracted to the gap. The attractive force results in strong confinement of the bead and thus a greatly suppressed Brownian motion, as shown by the red circles in Figure 8a. The full-width at half-maximum (FWHM) of the position distribution can be reduced by approximately 10 times from ~400 nm to ~40 nm.

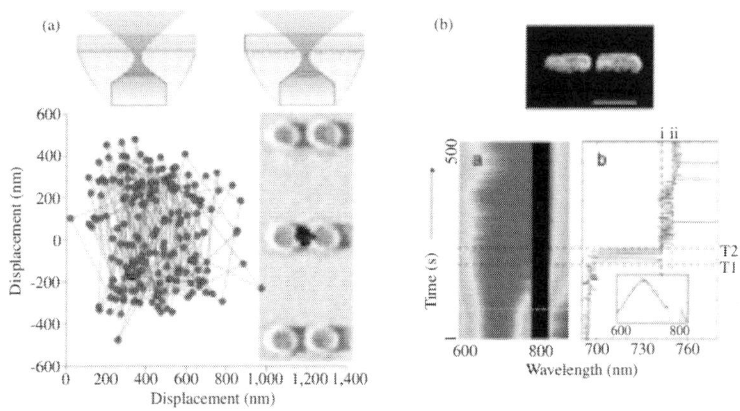


**Figure 9**. Near-field trapping of PS nanobeads (a) and gold nanoparticles (b) using a dimer of gold nanodisks (a) and nanocuboids (b).

When a metal nanoparticle is close to the gap of a dimer of gold nanocuboid, strong near-field coupling occurs between the metal nanoparticle and the nanoantenna, which produces extra near fields and enhances the trapping effect. Zhang reported the trapping of 10 nm GNPs at the gap of a gold dipole nanoantenna with a 5 to 30 nm gap sizeusing a slightly focused 808 nm laser. Individual trapping events are monitored in real time by the Rayleigh scattering of a single antenna. As shown in the lower panel of Figure 9b, when a GNP was trapped, a sudden red shift appeared due to the coupling between the particle and the antenna. The trapping behavior is also shown to be reversible, that is, by turning the trapping laser off, the trapped particle escapes from the gap.

## *Optical Forces on Metal Nanoparticles Trapped by a Focused Laser Beam*

Since optical force in plasmonic systems originates from the near fields, it exists not only in metal nanostructures that confine surface plasmons but also in metal thin films that support propagating surface plasmons. The difference is that in surface plasmon polariton (SPP) systems, the scattering force always pushes objects to move along the propagation. However, under counterbeam excitation or normal illumination, the propulsion can be canceled out and stable trapping is obtainable.

Figure 10 shows the measured force vectors (left) and magnitudes of a 2 μm dielectric bead as a function of the distance from a gold film (40 nm thickness) under 633 nm excitation at the plasmon resonance angle (~71° in this example). The z component is from the gradient of the evanescent field, and the x component is along the k-vector and is a result of plasmon propagation. It is clear that both components increase with decreasing distance, but the x component increases more rapidly than the z component, meaning that the dielectric bead will be pushed toward the plasmon propagation direction when it approaches the gold film. The right panel in Figure 9shows the optical force of beads of different sizes at constant surface separation (500 nm) from the film. Smaller particles experience larger forces, and the direction of the force vector is different for different particles. This effect could in principle be applied for optical sorting of nanoparticles.

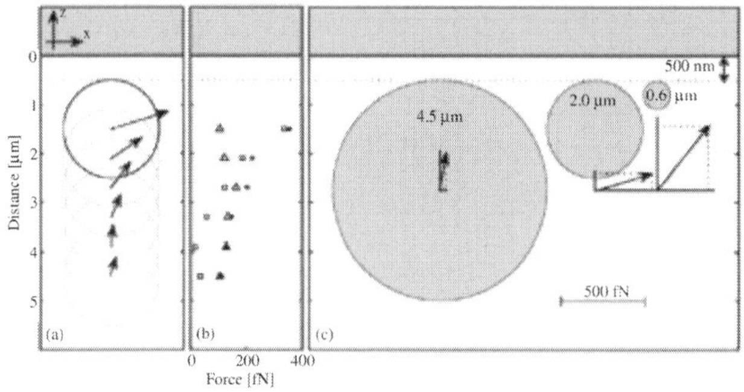


**Figure 10**. Optical force experienced by a dielectric bead close to a 40 nm thick gold thin film.

In the case of metallic nanoparticles, the propulsion is further enhanced due to the near-field coupling between the particle and the film. With counterpropagating SPPs excited by two incident beams at a specific incident angle, the propulsion from the two beams can cancel out to achieve stable trapping. Wang reported stable trapping of a dielectric bead (1 µm PS particle) on a gold stripe of 1 or 2 µm width. The experimental setup is shown in Figure 11a. Figure 11b shows the optical images of a PS bead (the bright spot in the center of the top-left image) propelled by a single laser beam. It is clearly seen that the particle is pushed to move toward the left side along the gold stripe. By using two counterbeams, the particle can be held stably at an equilibrium position in the middle determined by the intensities of the two incident beams. In Figure 10c, the position of a PS bead was recorded within 1 min at 30 measurements per second on a stripe of 1 µm width with equal laser intensity from both sides. It is seen that the particle is well confined along the x direction due to the electric field confinement, as shown in Figure 10d.

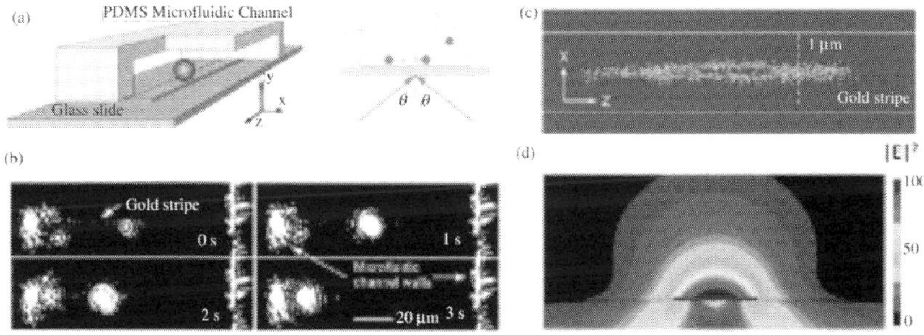


**Figure 11**. (a) Schematic of the experimental design. (b) Optical images of a PS bead pushed to move along the gold nanostripe. (c) Spatial positions of a PS bead on the gold nanostripe of 1 μm width illuminated from both left and right sides. (d) Simulated electric field component |E|2 of first-order quasi-TM mode on the gold stripe (1 μm wide and 40 nm thick).

## 5.3 PLASMON NANO-OPTICAL TWEEZERS

Conventional optical tweezers, formed at the diffraction-limited focus of a laser beam, have become a powerful and flexible tool for manipulating micrometre-sized objects. Extending optical trapping down to the nanometre scale would open unprecedented opportunities in many fields of science, where such nano-optical tweezers would allow the ultra-accurate positioning of single nano-objects. Among the possible strategies, the ability of metallic nanostructures to control light at the subwavelength scale can be exploited to engineer such nano-optical traps.

To understand the limitations faced by conventional optical tweezers1–4 when dealing with nanometre-sized specimens, we first consider an object of radius R much smaller than the wavelength of the incident trapping light (known as the Rayleigh regime). Reducing the size of the object causes two main effects that both work against stable trapping. First, the magnitude of the restoring force decreases abruptly (following an R3 law), which results in a shallower trapping well. Second, the damping of the trapped specimen decreases because of a reduction in viscous drag. Both effects alter the confinement of the specimen within the trap, as illustrated in Figure 12a–d. For a fixed average thermal energy kBT (where kB is the Boltzmann constant and T is the absolute temperature in the trap), this increasing delocalization allows the specimen to escape from the trap.

To compensate for this delocalization and prevent the specimen from escaping, one can either increase the trap confinement by further focusing the trapping laser light or augment the depth of the trapping potential by increasing the local laser intensity experienced by the specimen. The rule of thumb, established by Ashkin in his seminal work, states that stable trapping requires a potential depth of around 10kBT to compensate for stochastic kicks in the particle's Brownian motion.

Although diffraction prevents the confinement of propagating light beyond a fraction of its wavelength, recent advances in nanooptics have led to alternative techniques based on nano-optical tweezers that are capable of trapping and manipulating nanometre-sized objects. Indeed, unlike propagating fields, evanescent fields can be concentrated well beyond the diffraction limit. Among the different approaches, plasmon nano-optics, which exploits the surface plasmon resonances supported by metallic nanostructures, is particularly efficient in controlling light down to the nanometre scale. Plasmonic nanostructures can be engineered to efficiently couple to propagating light and concentrate it into highly localized and intense optical fields known as hot spots. By simultaneously increasing the confinement and augmenting the depth of the trapping potential, plasmonics opens up many new opportunities for optical trapping at the nanometre-scale.

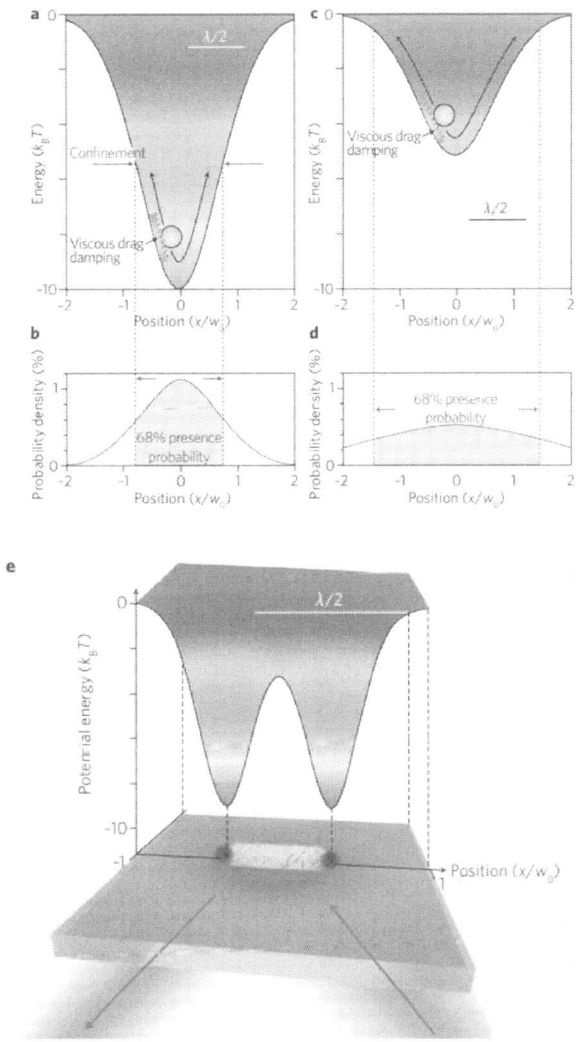


**Figure 12**. Conventional optical trapping versus surface plasmon trapping. a,b, Trapping potential (a) and associated position distribution (b) for a polystyrene sphere of radius R. c,d, Trapping potential (c) and associated position distribution (d) for a polystyrene sphere of radius 0.8R. The ratio between a and b is identical to that between c and d. Comparing a and b with c and d shows the dramatic impact that results from a size reduction of only 20%. e, Optical potential landscape resulting from the illumination of a gold nanostructure. The half-wavelength bar in panels a, c and e gives an approximate scale of reference.

### 5.3.1 Surface plasmons and optical forces

In the early years of research into near-field optics, several theoretical proposals were made regarding the use of evanescent fields for optical trapping. Novotny and Martin suggested exploiting the subwavelength light concentration at the extremity of a sharp metallic tip to create a nano-optical trap able to immobilize dielectric objects as small as a few tens of nanometres. At around the same time, Kawata. proposed a similar approach that produced a nano-optical trap using the confined light transmitted through a nanoaperture in an opaque metallic film. In both configurations, the interaction of the propagating incident laser with the nanostructure was predicted to create large k-vectors characteristic of evanescent waves, resulting in the subwavelength concentration of light. In other words, the illuminated nanostructure functions as a nanolens that is capable of concentrating light well beyond that of a high-numerical-aperture objective lens in a microscope. Interestingly, such nanostructure assisted formation of the optical potential makes it possible to create a complex near-field landscape from a single beam, including a large number of nanometre-sized traps for the parallel trapping of multiple nano-objects at predefined locations on a surface. Another attractive advantage of using a structured surface is its potential integrability into a compact platform without any need for bulky optical elements. However, the use of a fixed plasmonic pattern a priori limits the possibility of manipulating the object in three dimensions.

Experimental research on surface plasmon (SP)-based trapping was triggered by two pioneering studies into the enhanced force field at a homogeneous gold/water interface supporting a surface plasmon polariton (SPP). In 2006, Garcés-Chávez and co-workers reported the self-assembly of a large number of micrometre-sized dielectric beads within the region of a gold surface where an SPP was excited. The same year, Volpe used photonic force microscopy to probe the SPP force field, and measured a 40-fold enhancement to the force magnitude.

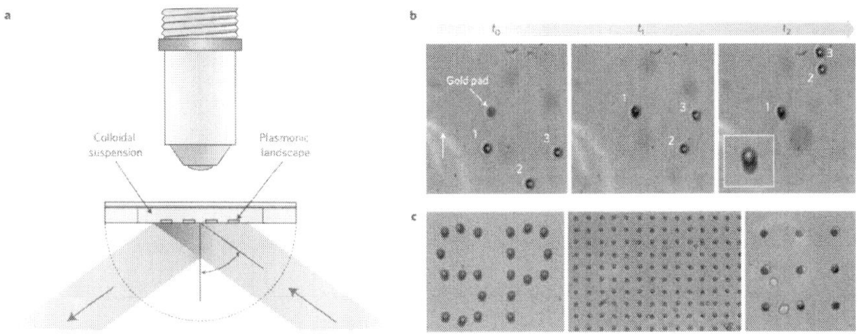


**Figure 13**. SP-based microtrapping. a, Schematic of the experimental set-up, in which a pattern of micrometre-sized gold disks is illuminated under the Kretschmann configuration through a glass prism. The black arrow gives the direction of the incident in-plane k-vector. b, Frame sequence illustrating the trapping of a 4.8 μm polystyrene bead in the forward position of a gold disk. c, Parallel trapping of polystyrene beads and yeast cells using a single laser beam31.

## 5.3.2 Plasmonic microtrapping

Although a homogeneous metallic film illuminated by an unfocused laser beam has a homogeneous optical force field, stable trapping at a predefined location on the film surface requires a confined trapping well, which can be achieved through metal patterning. The first experimental implementation of SP-based trapping was demonstrated using a glass surface decorated by micrometresized gold disks. In this experiment, the patterned surface was illuminated through a glass prism (known as the Kretschmann configuration) by an unfocused (~100 μm waist), linearly p-polarized near-infrared laser beam. The gold pattern was covered by a static fluidic chamber containing a diluted aqueous suspension of micrometre-sized polystyrene beads. When the direction of the incident light matched the SPP resonance angle, the light efficiently coupled to the SPP mode supported by the top gold/water interface. Owing to the combined effects of the boundaries and asymmetrical illumination, the plasmonic field was concentrated in the forward portion of the disk. No solution flow was used; it was the incident evanescent field that pushed the polystyrene beads at the glass–water interface along the incident in-plane k-vector, thus guiding them towards the trapping region of the surface. Figure 13b shows how a bead passing over one of the pads gets trapped in a forwards position, where the SPP field intensity is expected to be maximum. Trapping times of up to several hours were achieved for an incident laser intensity of 107 W $m^{-2}$ — around two orders of magnitude smaller than that required to trap a bead of the same size through conventional optical tweezers. To demonstrate that trapping is triggered by the pad plasmon resonance and rule out any sticking effects, the linear polarization of the illumination can be switched from p to s, for which light does not couple to the SPP. The resulting strong decrease in the near- field intensity at the gold surface makes the optical well shallower, hence releasing the bead. Because the pad is much smaller than the illumination spot, the experiment can be extended to an ensemble of pads to achieve, under the same illumination conditions, parallel trapping of a large number of beads. Beyond this proof-of-concept, parallel SP trapping provides potential new techniques for the manipulation of living cells by investigating their response to external stimuli.

In addition to being well-suited for integration on a chip, the in-plane geometry of SP traps provides them with some particularly attractive features. Similar to conventional three-dimensional optical tweezers, the trapping mechanism in SP tweezers relies on a suitable balance between scattering and gradient forces. Owing to the 2D geometry of SP traps, both the scattering (repulsive) and the gradient (attractive) forces are contained within the surface plane. However, in contrast with three-dimensional optical tweezers, both force contributions in SP trapping depend on illumination parameters such as the direction and polarization of the incident light. On the one hand, the magnitude of the scattering force increases as the angle decreases, but on the other hand, the incident angle also controls the gradient force component when scanning across the pad SPP resonance. Consequently, the trapping properties of SP traps can be tuned by changing the balance between the two competing force components. Such tuning has been exploited to selectively trap one particle size out of a mix of several sizes.

Although a purely electromagnetic model is sufficient to explain the trapping mechanism of SP tweezers, it is important to mention that optical forces may also coexist with thermally induced forces. In addition to optical gradients, coupling to SP resonances leads to local heating of the metal and heat dissipation to the surrounding fluid, which results in convection and thermophoresis. It was experimentally observed that the contribution of convection could be minimized by reducing the metal density (by exploiting well-separated gold disks) and using thin fluidic chambers with thicknesses smaller than 20 μm. So far there has been no clear data to quantify the respective contributions of optical and thermal effects.

### 5.3.3 Subwavelength trapping with plasmonic antennas

Although the simple disk geometry enables dielectric beads as small as 1 μm to be trapped by down-scaling the pad diameter accordingly, it fails for smaller sizes. At the heart of this failure is the change in nature of SPs when the disk becomes commensurable with or smaller than the SP wavelength. Owing to boundary effects, subwavelength gold structures support localized surface plasmon (LSP) resonances, which differ in several ways from SPPs in an extended film. Although gold pads measuring around 100 nm can be designed to be resonant in the near-infrared region of the spectrum, they are not suited for trapping because of their limited intensity and symmetry in the optical near field

Much higher control of plasmonic fields can be achieved through an alternative technique that exploits the strong electromagnetic coupling between several adjacent plasmonic nanostructures. Among the most interesting geometries, plasmonic antennas (also known as plasmonic dimers) have received much attention for their ability to concentrate propagating light well beyond the diffraction limit. Gap antennas usually consist of two identical metallic particles separated by a nanoscale dielectric gap. When the incident field is linearly polarized along the vector connecting the particles, capacitive effects lead to a confined and intense light spot within the gap region. Their huge potential for enhancing light–matter interactions at the nanoscale have raised the expectations of gap antennas in fields such as data processing, spectroscopy and sensing, and nonlinear optics.

The ability of gap antennas to concentrate laser light into very localized and intense hot spots makes them particularly attractive for creating optical traps orders of magnitude smaller than those achievable through the original pad approach, enabling SP trapping to be extended to the nanometre scale. Numerical calculations by Xu first predicted that hot spots in the gaps of metallic dimers could increase the residence time of single molecules in surface enhanced Raman spectroscopy experiments.

As a first experimental step in this direction, Grigorenko recently used gap antennas to assist conventional optical tweezers. In their experiment, a 200 nm polystyrene bead, trapped in a viscous medium at the tight focus of a 1,064 nm laser beam, was brought in close proximity to a surface patterned with pairs of gold cylinders designed to be resonant at the same wavelength.

As illustrated in Figure 14a, exposing the bead to the antenna hot spot modifies the overall optical potential seen by the bead, leading to a reduction in the trapping volume beyond the diffraction limit and thus providing an order-of-magnitude increase in bead confinement. In addition, the trapped object can be transferred between adjacent antennas by translating the laser beam, which allows the object to be discretely positioned.

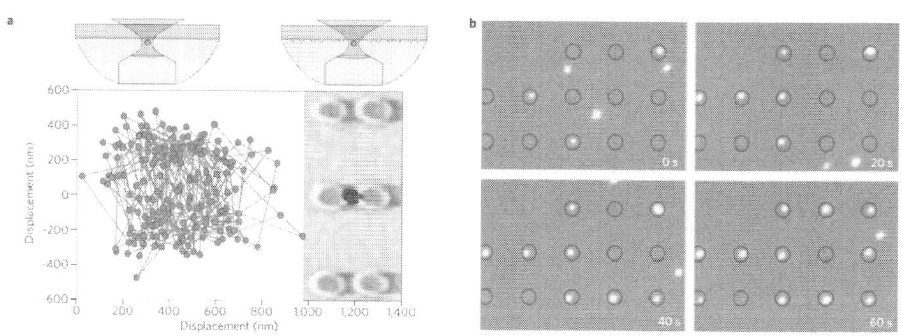

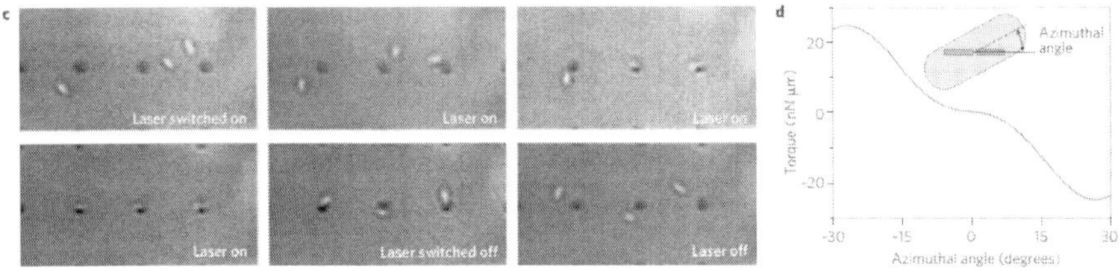


**Figure 14**. Antenna trapping.

Antennas were used as autonomous on-chip nano-optical traps capable of operating without the use of auxiliary conventional optical tweezers. We achieved parallel trapping of 200 nm polystyrene beads in aqueous solution using unfocused illumination and an array of gap antennas formed by adjacent gold nanobars. Such geometry supports a high-order resonance in the near-infrared that leads to a main hot spot within the gap and secondary hot spots along the bars and at their outer extremities. The applicability of this configuration to biological samples has also been demonstrated on living Escherichia coli bacteria. Interestingly, in addition to being efficiently trapped, the bacteria systematically align along the long axis of the antenna. This effect is due to the elongated shape of the bacteria, which enables them to exploit several trapping sites along the antenna until stabilized in a minimum-energy equilibrium position. Using a complex optical potential landscape formed by multiple nano-trapping sites to achieve further immobilization and controlled orientation is the nanoscale equivalent of multiple-beam trapping implemented with conventional optical tweezers for the accurate manipulation of cells. Stabilizing the specimen in the trap allows it to be optically inspected using techniques that require significant acquisition times, such as Raman spectroscopy.

### 5.3.4 Self-induced back-action trapping

Although enhanced and confined plasmonic fields allow dramatically strong field gradients and large local intensities to be achieved, previously discussed schemes suffer from a practical limitation that prevents trapping beyond a certain size. Indeed, one unavoidably reaches a limit above which the required intensity causes heat damage to the trapped specimen and/or the metallic structure. Even when not reaching such extreme limits, the thermal effects associated with intense fields may play an increasing role in the object's dynamics, and in some cases dominate over optical forces.

Extending optical trapping towards the nanometre regime therefore requires additional refinements to the trapping mechanism. The usual approach for optical trapping relies on a static picture, in which the potential well is optimized for a given object and remains constant over the experiment; that is, no modifications are applied to the potential well to account for the stochastic aspect of Brownian motion. Although the average kinetic energy of the object within the trap is given by $k_BT$, the instantaneous velocity follows a Maxwell–Boltzmann distribution, in which the object energy can occasionally significantly exceed the average value. Potential depths of around $10k_BT$ are typically used to compensate for these high-energy events; a depth of $k_BT$ would only hold the object until a high-energy event occurs.

Conversely, a dynamical approach, in which the trapping potential is dynamically reconfigured to compensate for high-energy events, would enable the required average trap depth to be strongly reduced. With this aim, external real-time feedback techniques have been applied to conventional optical tweezers to correct the trapping potential by adjusting either the laser intensity or the trap position. In practice, external feedback increases the effective stiffness of the trap, producing a force clamp and allowing the object to be further confined by adding an external source of damping50. Although this approach relaxes the requirements on the local intensity within the trap, thereby reducing the likelihood of optical damage, it relies on a complex auxiliary feedback set-up with a limited bandwidth, and is therefore sensitive to position measurement errors.

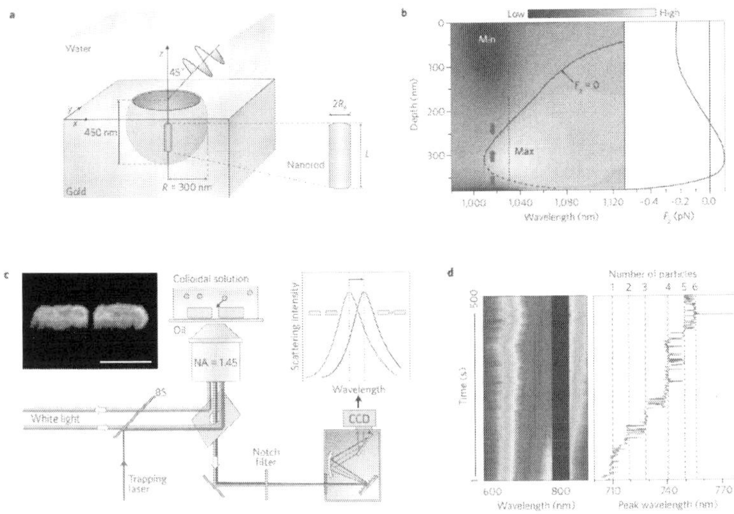


**Figure 15**. Plasmonic nanotrapping based on a trap–specimen interaction.

Alternatively, the extreme sensitivity of plasmonic nanostructures to their environment can be exploited to achieve some control over the features of an SP trap. The dependence of plasmon resonances to tiny changes of refractive index in the surrounding material has been extensively exploited to design efficient and compact sensors for gas, liquid and biomolecule detection. In the context of optical trapping, this sensitivity can have a destructive impact whenever the presence of the trapped object in the trapping site reduces the field gradient and/or the local intensity.

Conversely, plasmonic structures can also be devised such that the trapped object induces a constructive effect that favors trapping. García de Abajo and co-workers theoretically proposed exploiting the sensing ability of plasmonic nanostructures to achieve a reconfigurable nanotrap. In this configuration, a gold nanorod is trapped in the strong local field of a microvoid drilled in a gold film. Because the distribution and spectrum of the microvoid mode depend strongly on the nanorod position, the trapping equilibrium position can be tuned by changing the illumination wavelength. More recently, Zhang and co-workers reported efficient trapping of single gold nanoparticles within the gap of a gold antenna. Owing to their conductivity and high polarizability (about one order of magnitude higher than a polystyrene bead of the same size), gold nanoparticles have a strong impact on the gap antenna resonance by somehow shortcutting its two arms. Interestingly, the induced resonance shift allows the presence of the particle within the trap to be monitored.

Further steps involve fully exploiting the trap–specimen interaction to achieve an automatic feedback control that does not require any external monitoring or correcting. For this, one first aims to optimize the trapping efficiency by properly engineering the plasmonic mode such that the local intensity within the trap is maximized when the object is present. As a consequence, the momentum of the (plasmonic) photons interacting with the object experiences significant changes as the object moves in and out of the trap. Owing to momentum conservation, these changes create an additional dynamical force field that is by definition automatically synchronized with the object's dynamics. This 'self-induced backaction' (SIBA) was recently demonstrated using a nanoaperture in a metallic film.

SIBA trapping based on metallic nanoapertures relies on the high sensitivity of the aperture transmission to its dielectric environment. The presence of an object within the aperture, where the mode is confined, modifies the effective refractive index, hence red-shifting the transmission spectrum. Using a red-detuned laser such that the local field enhancement in the aperture gets stronger when the object is trapped leads to automatic positive back-action.

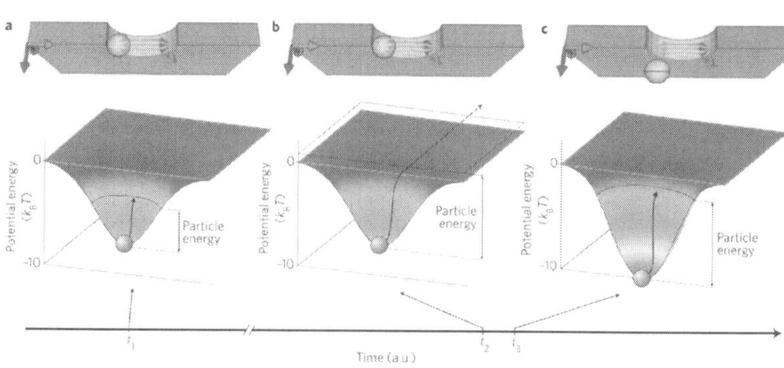

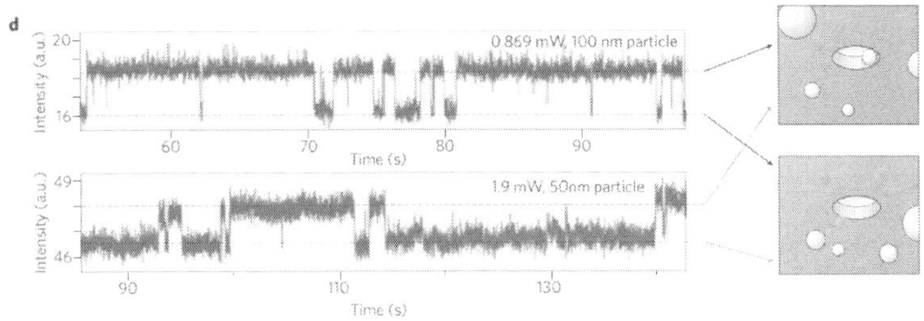


**Figure 16**. SIBA trapping.

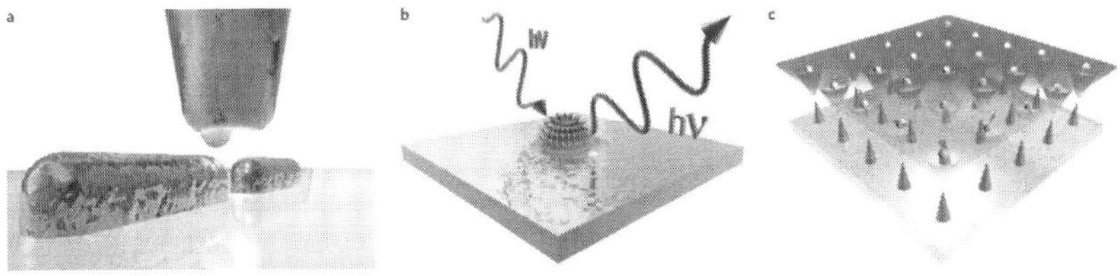


**Figure 17**. Future outlook and applications.

Whenever the total photon momentum travelling through the aperture decreases as the object escapes the trap, the SIBA force, pointing towards the equilibrium position, slows it down. This allows the incident laser intensity to be reduced to values associated with a potential depth of the order of kBT. The SIBA restoring force automatically increases the effective trapping potential depth during high-energy events to maintain the object within the trap. The SIBA effect has recently allowed efficient trapping of 100 nm and 50 nm polystyrene spheres with incident laser powers as low as 1 mW at a focus spot size of 1 $\mu m^2$. This approach shows a striking departure from previous techniques66–68. Although originally demonstrated with nanoapertures in a metallic film, the

SIBA approach relies on a more general phenomenon and could be applied to a large variety of geometries. In particular, recent efforts in the field of sensing based on the latest advances in plasmon nanooptics and metamaterials could significantly benefit from the optimization of SIBA traps. Careful optimization of the spatial overlap between the plasmonic mode and the trapped object, combined with an increase in the mode quality factor, should soon enable noninvasive single virus or protein trapping.

## REFERENCES

1. Acimovics, S., Kreuzer, M. P., Gonzalez, M. U. & Quidant, R. Plasmon nearfield coupling in metal dimers as a step toward single-molecule sensing. ACS Nano 3, 1231–1237 (2009).
2. Aizpurua, J. et al. Optical properties of coupled metallic nanorods for fieldenhanced spectroscopy. Phys. Rev. B 71, 235420 (2005).
3. Barnes, W. L., Dereux, A. & Ebbesen, T. W. Surface plasmon subwavelength optics. Nature 424, 824–830 (2003).
4. Chaumet, P., Rahmani, A. & Nieto-Vesperinas, M. Optical trapping and manipulation of nano-objects with an apertureless probe. Phys. Rev. Lett. 88, 123601 (2002).
5. Cizmar, T., Garcés-Chávez, V., Dholakia, K. & Zemánek, P. Optical conveyor belt for delivery of submicron objects. Appl. Phys. Lett. 86, 174101 (2005).
6. Dholakia, K. & Reece, P. Optical micromanipulation takes hold. Nano Today 1, 18–27 (2006).
7. Garcés-Chávez, V. et al. Extended organization of colloidal microparticles by surface plasmon polariton excitation. Phys. Rev. B 73, 085417 (2006).
8. Garcia-Parajo, M. F. Optical antennas focus in on biology. Nature Photon. 2, 201–203 (2008).
9. Grier, D. A revolution in optical manipulation. Nature 424, 810–816 (2003).
10. Grigorenko, A. N., Roberts, N. W., Dickinson, M. R. & Zhang, Y. Nanometric optical tweezers based on nanostructured substrates. Nature Photon. 2, 365–370 (2008).
11. Gu, M., Haumonte, J. B., Micheau Y., Chon, J. W. M. & Gan, X. Laser trapping and manipulation under focused evanescent wave illumination. Appl. Phys. Lett. 84, 4236–4238 (2004).
12. Huang, L., Maerkl, S. J. & Martin, O. J. F. Integration of plasmonic trapping in a microfluidic environment. Opt. Express 17, 6018–6024 (2009).
13. Kim, S. et al. High-harmonic generation by resonant plasmon

field enhancement. Nature 453, 757–760 (2008).
14. Kinkhabwala, A. et al. Large single-molecule fluorescence enhancements produced by a bowtie nanoantenna. Nature Photon. 3, 654–657 (2009).
15. Li, K., Stockman, M. I. & Bergman, D. J. Self-similar chain of metal nanospheres as an efficient nanolens. Phys. Rev. Lett. 91, 227402 (2003).
16. Mühlschlegel, P., Eisler, H. J., Martin, O. J. F., Hecht, B. & Pohl, D. W. Resonant optical antennas. Science 308, 1607–1609 (2006).
17. Neuman K. C. & Block S. M. Optical trapping. Rev. Sci. Instrum. 75, 2787–2809 (2004).
18. Nieto-Vesperinas, M., Chaumet, P. & Rahmani, A. Near field photonic forces. Phil. Trans. R. Soc. A 362, 719–737 (2004).
19. Ozbay, E. Plasmonics: merging photonics and electronics at nanoscale dimensions. Science 311, 189–193 (2006).
20. Quidant, R., Petrov, D. & Badenes, G. Radiation forces on a Rayleigh dielectric sphere in a patterned optical near field. Opt. Lett. 30, 1009–1011 (2005).
21. Righini, M. et al. Nano-optical trapping of Rayleigh particles and Escherichia coli bacteria with resonant optical antennas. Nano Lett. 9, 3387–3391 (2009).
22. Righini, M., Volpe, G., Girard, C., Petrov, D. & Quidant, R. Surface plasmon optical tweezers: tunable optical manipulation in the femtonewton range. Phys. Rev. Lett. 100, 183604 (2008).
23. Righini, M., Zelenina, A. S., Girard, C. & Quidant, R. Parallel and selective trapping in a patterned plasmonic landscape. Nature Phys. 3, 477–480 (2007).
24. Schuck, P. J., Fromm, D. P., Sundaramurthy, A., Kino, G. S. & Moerner, W. E. Improving the mismatch between light and nanoscale objects with gold bowtie nanoantennas. Phys. Rev. Lett. 94, 017402 (2005).
25. Schuller, J. A. et al. Plasmonics for extreme light concentration and manipulation. Nature Mater. 9, 193–204 (2010).

26. Stockman, M. I. Nanofocusing of optical energy in tapered plasmonic waveguides. Phys. Rev. Lett. 93, 137404 (2004).
27. Tang, L. et al. Nanometre-scale germanium photodetector enhanced by a nearinfrared dipole antenna. Nature Photon. 2, 226–229 (2008)
28. Volpe, G., Quidant, R., Badenes, G. & Petrov, D. Surface plasmon radiation forces. Phys. Rev. Lett. 96, 238101 (2006).
29. Xu, H. & Käll, M. Surface-plasmon-enhanced optical forces in silver nanoaggregates. Phys. Rev. Lett. 89, 246802 (2002).

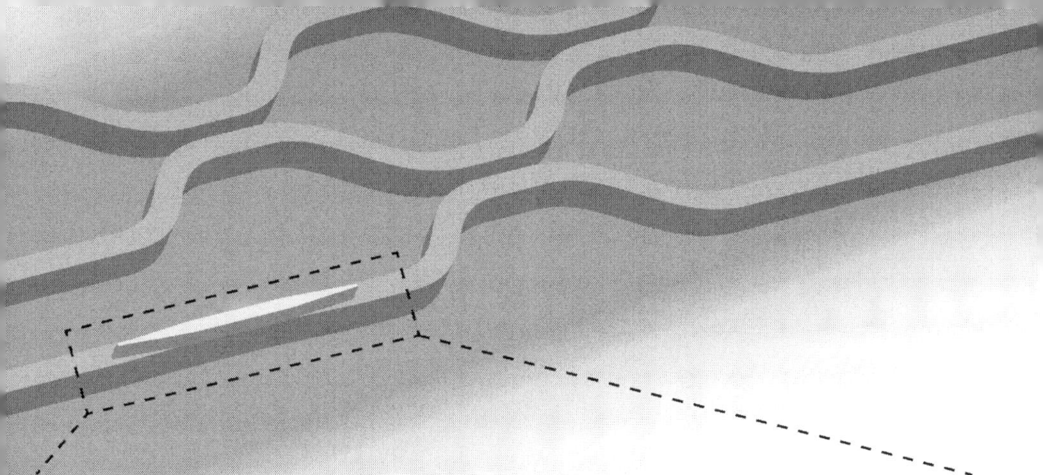

CHAPTER 6

# PLASMONIC NANOWIRE WAVEGUIDES AND CIRCUITS

## INTRODUCTION

As typical one-dimensional nanostructures for waveguiding tightly confined optical fields beyond the diffraction limit, metal nanowires have been used as versatile nanoscale building blocks for functional plasmonic and photonic structures and devices. Metal nanowires, especially those fabricated by bottom-up synthesis such as Ag and Au nanowires, usually exhibit excellent diameter uniformity and surface smoothness with diameters down to tens of nanometers, which offers great opportunities for plasmonic waveguiding of optical fields with deep-subwavelength confinement, coherence maintenance and low scattering losses. Based on nanowire plasmonic waveguides, a variety of applications ranging from plasmonic couplers, interferometers, resonators to photon emitters have been reported in recent years.

The fast development of plasmonics have greatly advanced our understanding to the abundant phenomena related to surface plamon polaritons (SPPs) and improved our ability to manipulate light at the nanometer scale. With tightly confined local field, SPPs can be transmitted in waveguides of subwavelength dimensions. Nanophotonic circuits built with plasmonic elements can be scaled down to dimensions compatible with semiconductor-based nanoelectronic circuits, which provides a potential solution for the next-generation information technology. Different structures have been explored as plasmonic waveguides for potential integration applications.

## 6.1 EXCITATION AND DETECTION OF PROPAGATING SPPS

Surface plasmon polaritons (SPPs) are electromagnetic waves that travel along a metal–dielectric or metal–air interface, practically in the infrared or visible-frequency. The term "surface plasmon polariton" explains that the wave involves both charge motion in the metal ("surface plasmon") and electromagnetic waves in the air or dielectric ("polariton").

They are a type of surface wave, guided along the interface in much the same way that light can be guided by an optical fiber. SPPs are shorter in wavelength than the incident light (photons). Hence, SPPs can have tighter spatial confinement and higher local field intensity. Perpendicular to the interface, they have subwavelength-scale confinement. An SPP will propagate along the interface until its energy is lost either to absorption in the metal or scattering into other directions (such as into free space).

Application of SPPs enables subwavelength optics in microscopy and lithography beyond the diffraction limit. It also enables the first steady-state micro-mechanical measurement of a fundamental property of light itself: the momentum of a photon in a dielectric medium. Other applications are photonic data storage, light generation, and bio-photonics.

### 6.1.1 Excitation of Surface Plasmon Polaritons

The SPP dispersion curve lies entirely below that of free space light in the dielectric, such that β > k. Direct excitation of SPPs by light beams is thus not possible unless special techniques to achieve phase matching are employed. These will be discussed here.

*Prism coupling*

This coupling scheme, also known as attenuated total internal reflection, involves the coupling of the SPPs to the evanescent electro-magnetic field that is formed upon total internal reflection of a light beam at a surface in an optically dense medium. Two different geometries for prism coupling are possible, see figure 1.

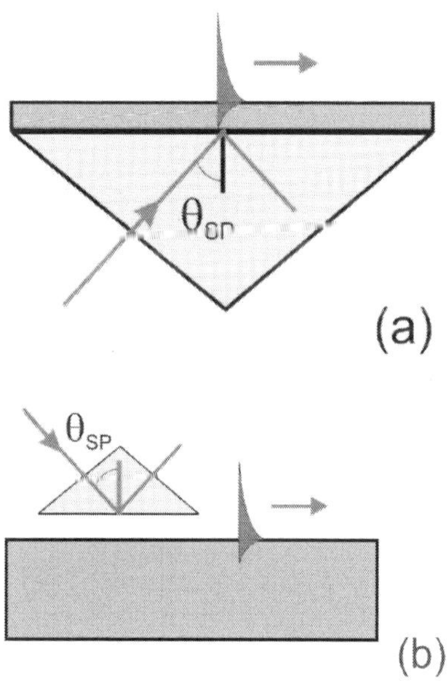


**Figure 1:** Prism coupling: (a) Kretschmann, (b) Otto configuration.

In the Kretschmann configuration, the metal film is evaporated on top of a glass prism. The film is illuminated through the dielectric prism at an angle of incidence greater than the angle of total internal reflection. The wavevector of light is increased in the optically dense medium. At a certain angle θ of incidence where the in-plane component of the photon wavevector in the prism coincides with the SPP wavevector on an air–metal surface, resonant light tunnelling through the metal film occurs and light is coupled to the surface polaritons:

$$\beta = \frac{\omega}{c}\sqrt{\epsilon_{\text{prism}}}\sin\theta$$

Under these resonant conditions, a sharp minimum is observed in the reflectivity from the prism interface as light can be coupled to SPPs with almost 100% efficiency. With the increase in the metal film thickness the efficiency of the SPP excitation decreases as the tunnelling distance increases. SPP on an interface between the prism and metal cannot be excited in this geometry as the wavevector of SPP at this interface is greater than the photon wavevector in the prism at all angles of incidence. To be able to excite SPP on the internal metal interface, an additional dielectric layer with a refractive index smaller that of the prism should be deposited between the prism and the metal film. In such a two-layer geometry, the photon tunnelling through this additional dielectric layer can provide resonant excitation of SPP on the inner interface. Thus, both SPP modes (on the surface and the interface) can be excited in such a configuration at different angles of illumination. For thick metal films (or surfaces of bulk metal), for which the Kretschmann configuration cannot be used, SPP can be excited in the Otto configuration. Here, the prism where total internal reflection occurs is placed close to the metal surface, so that photon tunnelling occurs through the air gap between the prism and the surface. The resonant conditions are analogous to those in the Kretschmann configuration. This configuration is also preferable when direct contact with the metal surface is undesirable, e.g. for studying the surface quality.

### *Grating coupling*

The mismatch in wavevector between the in-plane momentum $k_x = k \sin \theta$ of impinging photons and $\beta$ can also be overcome by using diffraction effects at a grating pattern on the metal surface. For a one-dimensional grating of grooves with lattice constant a, as depicted in figure 2, phase-matching takes place whenever the condition

$$\beta = k \sin\theta \pm nG$$

is fulfilled, where G = 2π/a is the reciprocal vector of the grating and n = 1, 2, 3, . . .. As with prism coupling, excitation of SPPs is detected as a minimum in the reflected light. The reverse process can also take place, when SPPs propagating along a modulated surface can couple to light and thus radiate (see figure 3).

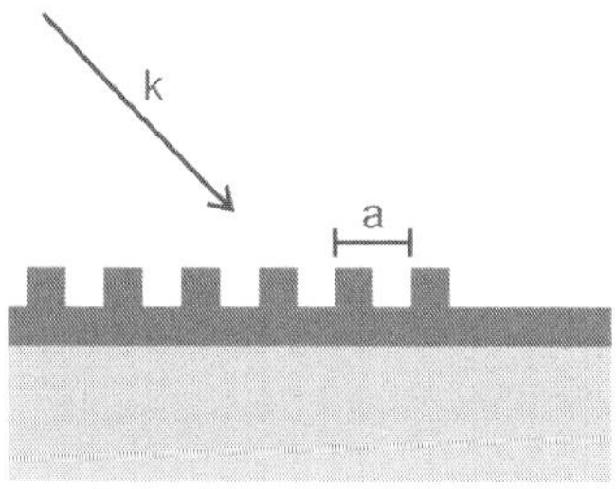


**Figure 2**: Grating coupling of a light with wave vector k impinging on a metal grating surface of period a.

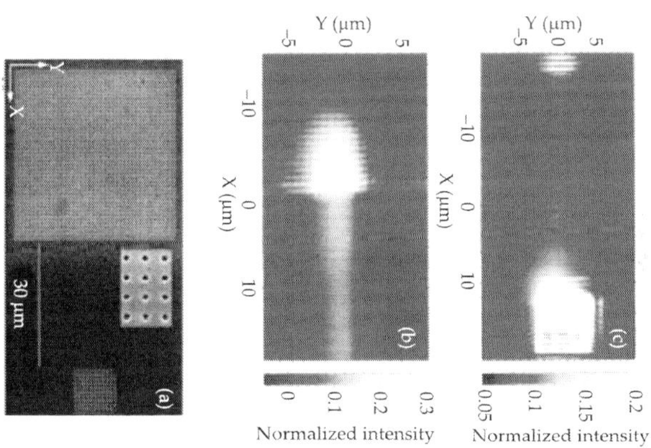


**Figure 3:** Left: SEM image of the pattern. 2 arrays with a period of 760 nm are separated by 30 mm. The hole diameter is 250 nm. The small array at the bottom is the source array whereas the big array at the top is not illuminated by the laser light and acts as a probe array. Inset: zoom in on the holes of one array. Right: Near-field optical image of the pattern when the laser is focused on the small array and the electric field is along the x direction.

For one-dimensional gratings, significant changes to the SPP dispersion relations occur if the gratings are sufficiently deep so that the modulation cannot longer be treated as a small perturbation. An SPP photonic bandgap is then formed when the period is equal to half the wavelength of the SPP. Just as for electron waves in crystalline solids, there are two SPP standing wave solutions, each with the same wavelength but, owing to their different field and surface charge distributions, they are of different frequencies (see figure 4(a)). The upper frequency solution $\omega+$ is of higher energy because of the greater distance between the surface charges and the greater distortion of the field. SPP modes with frequencies between the two band edges cannot propagate, and so this frequency interval is known as a stop gap.

## *Excitation using highly focused optical beams*

As a variant of the prism coupling technique, a microscope objective of high numerical aperture can be used for SPP excitation. An oil immersion objective is brought into contact with the glass substrate of a thin metal film via a layer of index matched immersion oil. The high N.A. of the objective ensures a large angular spread of the focused excitation beam, including angles $\theta > \theta_c$ greater than the critical angle for total internal reflection. In this way, wave vectors $kx = \beta$ are available for phase-matching to SPPs at the metal/air interface.

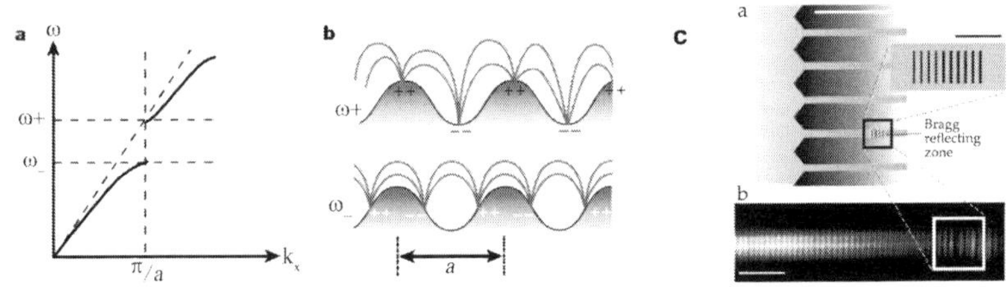


**Figure 4:** (a) SPP dispersion relation of a periodically modulated surface. (b) Schematic representation of the SPP modes near the band edges. (c) Realization of a SPP Bragg reflector.

Off-axis entrance of the excitation beam can further ensure an intensity distribution preferentially around $\theta_{SPP}$ thus reducing the amount of directly transmitted/reflected light. The highly focused beam also allows for localized excitation within the diffraction limit. This scheme is especially convenient for the excitation of a continuum of SPPs at different frequencies. And the determination of their propagation lengths.

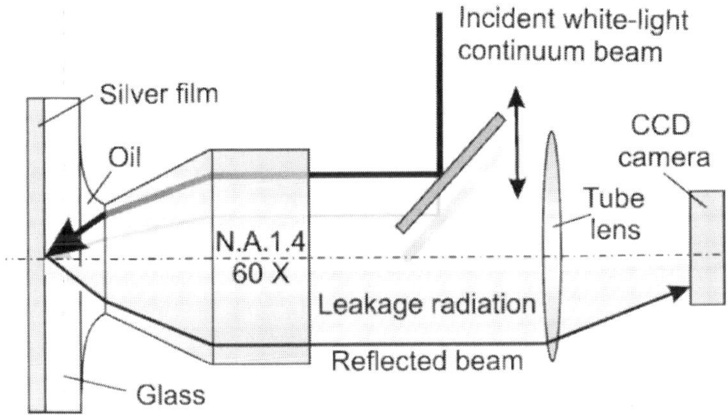


**Figure 5:** Schematic of the excitation of a white-light continuum of SPPs and their observation through an index-matched oil-immersion lens.

### *Near-field excitation*

Near-field applications in SPP studies brought forth a new technique which provides the possibility to excite SPP locally with sub-wavelength resolution. Additionally, the effect of surface roughness on the SPP propagation and the scattering at individual surface defects can be studied with high spatial resolution.

Using illumination through a SNOM fiber tip, circular SPP waves can be locally launched at the surface. This configuration (see figure 6) can be treated as either a diffraction or tunnelling mechanism of the SPP excitation. In the former case, one can consider a near-field coupling of the light diffracted on the sub-wavelength aperture of the fiber tip in SPPs. In the latter case, because mainly the evanescent field components are generated at the aperture, one can describe the process as the photon tunnelling taking place from the fiber tip to the metal surface. This technique is somewhat analogous to the Otto configuration but allows local (with sub-wavelength precision) excitation of SPPs at the SNOM tip position.

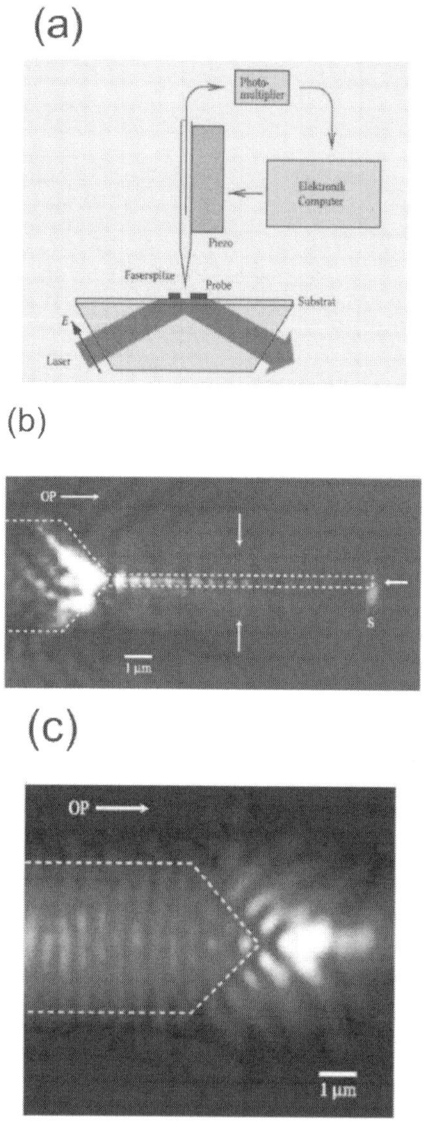


**Figure 6:** (a) Principle of the detection of SPPs (excited via the prism) with a SNOM probe. (b) and (c) SNOM images, propagation of SPPs along gold nanowires (wavelength 633 nm).

### *Coupling to integrated photonic elements*

While the excitation schemes above are interesting for investigation of SPP propagation, practical applications of SPPs will require compact and efficient coupling schemes in integrated photonic circuits. Preferably, the plasmonic components should allow efficient matching with conventional dielectric waveguides and fibers.

One such coupling scheme is end-fire coupling, in which a free-space optical beam is focused on the end-facet of the desired waveguide. This scheme operates via matching the spatial field distribution of the waveguide as much as possible by adjusting the beam width. This can be further miniaturized by directly connecting an optical fiber to the end-faced of the plasmon waveguide (see figure 7).

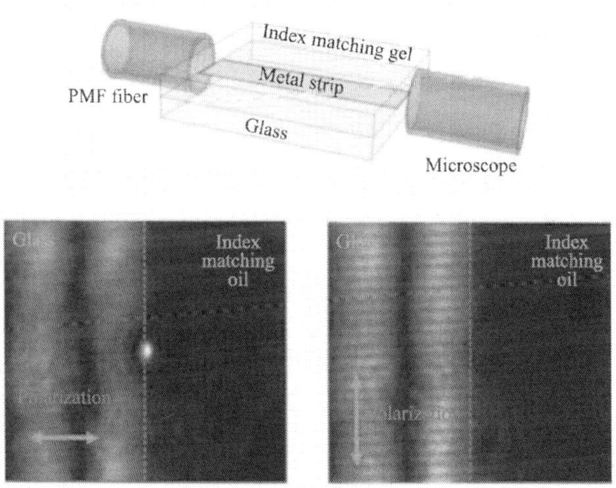


**Figure 7:** A polarization maintaining fiber coupled to a HeNe laser illuminates one end of the strip and the microscope collects the light scattered at the other end. The strip output lights up when the electric field is parallel to the short axis of the strip. This result is as expected because the SPP modes are mostly TM polarized.

A convenient interfacing scheme makes use of optical fiber tapers brought into the immediate vicinity of the waveguide to achieve phase-matched power transfer via evanescent coupling (see figure 8). Fiber tapers are standard glass optical fibers that have been heated and stretched to a diameter that is at or below the wavelength of light, at which point the evanescent field of the guided mode extends into the surrounding air and allows the taper to function as a near-field optic. Similar approaches are used to couple light to microcavity structures such as microdisks.

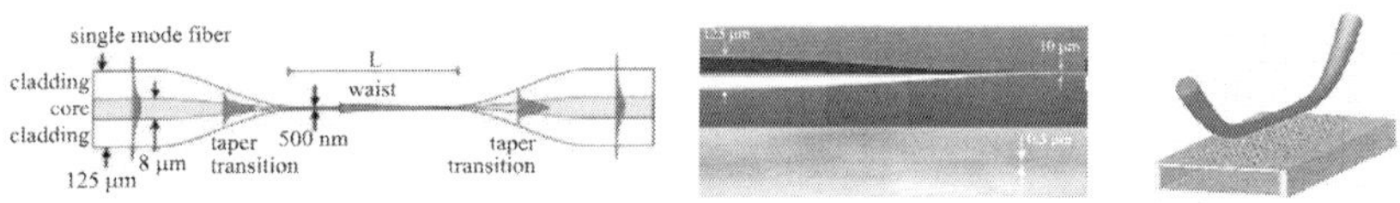


**Figure 8:** Left and center: Fiber tapers. Right: Fiber taper coupled to an SPP nanoparticle waveguide.

## 6.1.2 Coupling to Surface Plasmon Polaritons

The interaction between the electromagnetic field and the surface charges results in an increase in the SPP momentum compared with that of the free space according to the SPP dispersion relation), which gives rise to the momentum mismatch between light and SPP. The momentum of light and SPP can be matched using different coupler configurations such as prism couplers, grating couplers, fibre and waveguide couplers.

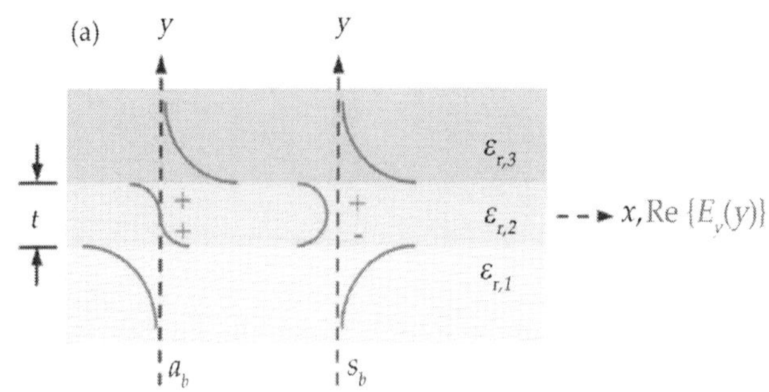

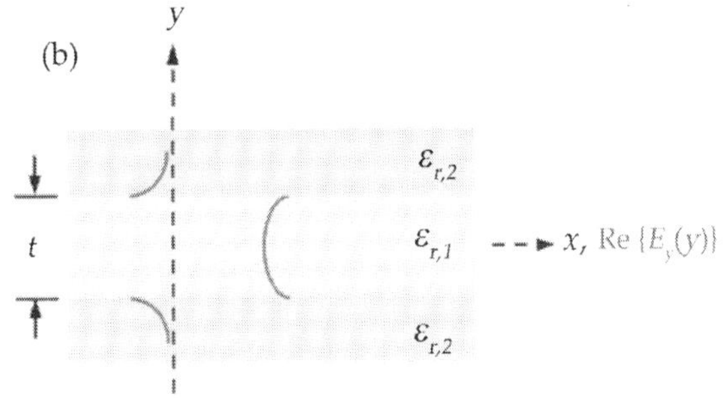


**Figure 9.** (a) Metal slab ($\varepsilon_{r,2}$) of thickness t bounded by semi-infinite dielectrics ($\varepsilon_{r,1}$, $\varepsilon_{r,3}$) supporting two bound SPP modes ($a_b$, $s_b$). (b) Dielectric slab ($\varepsilon_{r,1}$) of thickness t bounded by semi-infinite metals ($\varepsilon_{r,2}$) supporting a symmetric bound mode. The distribution of the main transverse electric field component of the modes is shown as red curves over the cross-section of each structure.

The metal film is illuminated through the dielectric prism at an angle of incidence greater than the critical angle or the total internal reflection angle in the Kretschmann configuration (figure 10(a)). The light wavevector increases in the optically dense medium at a certain incident angle larger than the critical angle. The in-plane component of the light wavevector in the prism coincides with the SPP wavevector on an air–metal surface, which gives rise to the light tunnelling through the metal film, and as a result, the light is coupled to the SPP. Furthermore, the efficiency of SPP excitation decreases with increasing film thickness due to the increase in the tunnelling distance. Figure 10(b) shows a two-layer geometry (a dielectric layer with a refractive index smaller than that of the prism is deposited between the prism and the metal film), and using this geometry, the light tunnelling through this dielectric layer can excite SPPs on the internal metal interface at different incident angles. For a thick metal film, Kretschmann configuration cannot be used to excite SPPs; Otto configuration can overcome the problem by placing the prism close to the metal surface (figure 10(c)). In this case, light tunnelling occurs through the air gap between the prism and the surface. On the other hand, the diffraction effects by a diffraction grating can satisfy the momentum matching conditions (figures 10(e) and (f)). The diffracted orders from the periodically corrugated metal–dielectric interface exhibit larger wave vectors in magnitude than those of the incident light, and as a result, the diffraction of light by the metal grating will be coupled to the SPP.

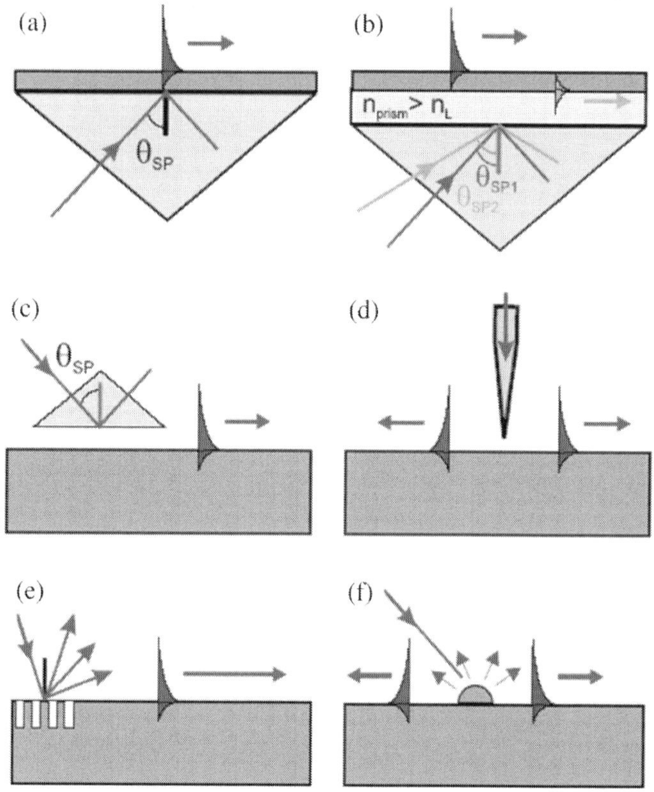


**Figure 10.** SPP excitation configurations: (a) Kretschmann geometry, (b) two-layer Kretschmann geometry, (c) Otto geometry, (d) excitation with an SNOM probe, (e) diffraction on a grating and (f ) diffraction on surface features.

Considering the principle of SPP excitation based on total internal reflection with a prism coupler, the coupling prism can be replaced by a waveguide layer of a fibre core; therefore, the light can be coupled to the SPP using a fibre coupler. The cladding of the fibre is removed and it is coated with a metal layer, which is surrounded by a dielectric sensed medium. This coupling of the evanescent field with SPP strongly depends upon the light wavelength, fibre parameters, fibre geometry and metal layer properties . For example, the coupling mechanism will be different for single-mode and multi-mode optical fibres due to different mode transmission properties. In particular, a straight fibre and a tapered fibre exhibit different strengths of light coupling because these fibres show different penetration depths of the evanescent field due to different geometrical configurations. The advantages of optical fibre couplers are mainly that they are miniaturized and flexible. In addition, the light can also be coupled to the SPP using an optical waveguide coupler. The process of exciting SPPs using a optical waveguide is similar to that in the Kretschmann configuration (figure 10(b)). The light is guided by either a single or multilayer (slab or channel) waveguide into a region with a thin metal overlayer, and the light penetrates evanescently through the metal layer. If the SPP and the guided mode can be phase-matched, the incident light excites the SPP at the outer interface of the metal layer.

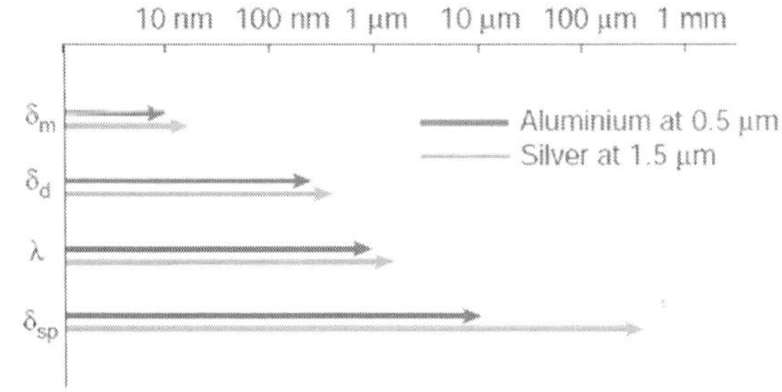


**Figure 11.** Comparison of the propagation lengths of SPPs between silver and aluminium.

## 6.1.3 Propagation of surface plasmon polaritons

Once light has been converted into an SPP mode on a flat metal surface it will propagate, but will gradually attenuate owing to losses arising from absorption in the metal. The propagation length of the SPP is thereby limited by the imaginary part of the complex SPP wavevector $k_{spp}$ due to the internal damping (ohmic losses), where $k_{spp} = k_{sppr} + ik_{sppi}$. Based on the SPP dispersion relation, the propagation length of SPP $\delta_{spp}$ is given by

$$\delta_{spp} = \frac{1}{2k_{spp\,i}} = \frac{\lambda}{2\pi}\left(\frac{\varepsilon_{mr}+\varepsilon_d}{\varepsilon_{mr}\varepsilon_d}\right)^{3/2}\frac{\varepsilon_{mr}^2}{\varepsilon_{mi}} \qquad (1)$$

where $\varepsilon_{mr}$ and $\varepsilon_{mi}$ are the real and imaginary parts of the dielectric function of the metal, respectively, that is, $\varepsilon_m = \varepsilon_{mr} + i\varepsilon_{mi}$. where $\delta_{spp}$ is the distance after which the SPP intensity decreases to 1/e of its starting value. The propagation length is usually dependent on the dielectric constant of the metal and the incident wavelength based on equation (1). For example, for a relatively absorbing metal such as aluminium the propagation length is 2μm at a wavelength of 500 nm, while for a low loss metal such as silver, at the same wavelength it is up to 20μm (figure 11). In addition, for a longer incident wavelength, such as the near-infrared telecom wavelength 1.55μm, the propagation length of silver increases towards 1 mm.

## 6.2FUNDAMENTALPROPERTIESOFSPPSIN METAL NANOWIRES

Various nano-structures have been exploited to confine and propagate SPPs such as thin metal films, micrometer-wide stripes, nanowires, chains or arrays of nanoparticles, nanoholes, nanogaps, nanogrooves and wedges, hybrid waveguides and so on. The propagation of surface plasmons was first studied in thin metal films. With the development of scanning near-field optical microscopy, it was soon realized that the propagation of SPPs in thin films was affected by structural defects present at the film surface. Artificially creating these defects at the surface opened the door to what was called "plasmon optics": manipulating a surface plasmon beam very much like standard optical elements are shaping a laser beam.

Spreading of the surface plasmon in the film naturally occurs in both in-plane directions with an extent that depends on the type of excitation, but also on the quality of the film. In order to confine the SPP mode in one direction, the film width was reduced laterally to form stripe metal waveguides. Propagation of the SPP is restricted to the stripe since it does not exist outside the metal. Mode formation in stripes is understood as total internal reflection of the SPP and edge interaction. Stripe waveguides are however suffering from a strong limitation detrimental for realizing dense plasmonic network. As the width of the stripe is reduced to increase lateral confinement, the SPP mode runs into a cutoff and becomes leaky for width <2 to 3 μm.

Benefiting from near-field coupling and well defined dipolar surface plasmon resonances, Stefan Maier and his co-workers used chains of nanoparticles to increase SPP confinement and overcome the cutoff limit of stripe waveguide. An example of an array of such discrete structures is illustrated in Fig. 12(a). However, ohmic and scattering losses are dominant in nanoparticles and the propagation was restricted to a few hundred nanometers. Other plasmon waveguides such as nanoholes (Fig. 12(b)), nanogrooves (Fig. 12(c)) and nanogaps (Fig. 12(d)) can also provide strong localization mechanisms while limiting extrinsic losses such as scattering. Like any other plasmonic waveguides, these approaches are plagued by the trade-off existing between the waveguide size and propagation loss. Because metals at optical or near-infrared wavelengths have a field skin depth comparable to the thickness of the structure, confined modes are associated with large intrinsic losses. The propagation distance is therefore reduced compared to modes spreading more outside the metal. An hybrid waveguide consisting of a silicon waveguide and a silver nanowire in Fig. 12(e) was proposed as a solution for this trade-off. These waveguides are usually designed for longer wavelength.

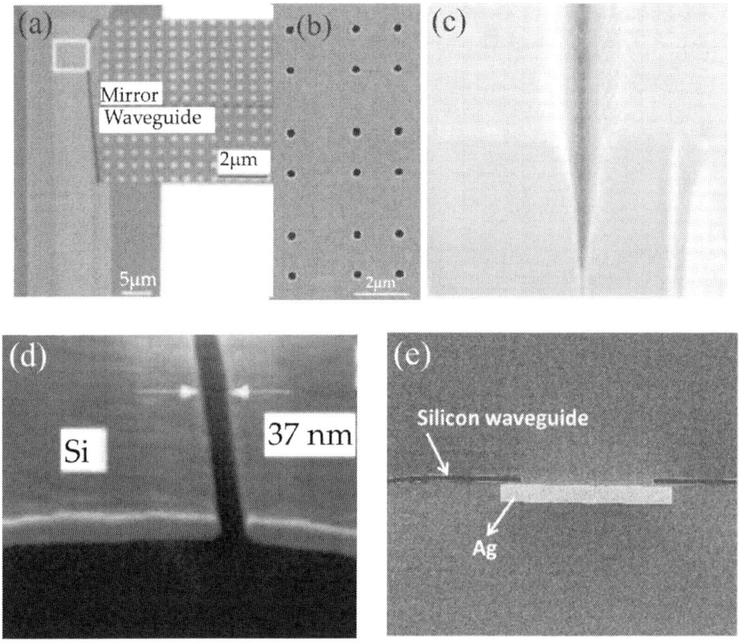


**Figure 12:** Different type of plasmon waveguides: arrays of Au nanoparticles (a), nanoholes (b) , nanogroove (c), nanogap (d), and an hybrid waveguide consisting of a silicon waveguide and a Ag nanowire (e).

Compared to the available plasmon waveguides, metal nanowires are capable of supporting SPPs propagating along one direction while localizing SPPs in the other two. Unlike metal films and metal stripes, there is no cutoff for the fundamental mode in metal nanowires. Furthermore, the single-crystal nature of chemically synthesized metal nanowires mitigates propagation losses. A good trade-off can thus be obtained between waveguide size and plasmon decay length for optimized nanowires.

### 6.2.1 Properties of SPP propagating in metal nanowires

*Dispersion relation in nanowires*

The electromagnetic field of a surface plasmon is the solution of Maxwell's equations in each medium taking into account boundary conditions. The derivation of the SPP dispersion relation in metal film is obtained by considering a system consisting of a dielectric medium with a real and positive dielectric constant $\epsilon_d$ and a metal with complex dielectric constant $\epsilon_m = \epsilon'_m + i\epsilon''_m$. By solving Maxwell's equations and the associated boundary conditions, the dispersion relation of SPP in thin film can be obtained. However, when SPPs propagate in one-dimensional nanowire waveguides, the geometry of the systems imposes a different dispersion.

The derivation of plasmon dispersion in nanowires was first deduced by considering one-dimensional negative dielectric (ND) pin model with cylindrical cross-section shown in Fig. 13(a). Since a metal is a negative dielectric in the near-infrared range, the model treats the metal as an optical waveguide. Cylindrical coordinates were chosen in this model. The optical wave was assumed to propagate in the z direction with a phase constant β and an angular frequency ω. The lowest order (0th) TM mode was considered. By solving Maxwell's equations, the electric and magnetic field in the two different dielectric medium $\epsilon_1$ and $\epsilon_2$ can be obtained. From the boundary conditions for $E_z$ and $H_\varphi$ at r = a, where r is the radius direction of the cylindrical coordinates, $\alpha$ is the radius of the core, the characteristic equation of the 0th-order TM mode reads as follow:

$$\frac{\gamma_2 I_1(\gamma_1 a) K_0(\gamma_2 a)}{\gamma_1 I_0(\gamma_1 a) K_1(\gamma_2 a)} = -\frac{\epsilon_2}{\epsilon_1},$$

where $I_j$ and $K_j$ (j = 1, 2) are the jth-order modified Bessel functions. $\gamma_j$ is defined as

$$\gamma_j = (\beta^2 - \epsilon_j \mu_0 \omega^2)^{1/2},$$

where $\mu_0$ is the permeability of free space and $\varepsilon_j$ is the dielectric constant. In the case of a metal, ϙ1 can be regarded as the dielectric constant of metal $\varepsilon_m$ and $\varepsilon_2$ can be regarded as the dielectric constant of surrounding medium $\varepsilon_d$.

Following the pioneering work of Takahara, the SPP dispersion in cylindrical nanowires were extensively studied form theoretical considerations and from experimental data. Near-field spectroscopy and far-field extinction spectroscopy were used to obtain the dispersion relation of SPPs in metal nanowires. Figure 13(b) displays such a dispersion for an array of nanowires measured by extinction spectroscopy. It shows clearly the dispersion differences for light in vacuum, glass and SPP supported by a flat interface and as a mode of a Ag nanowire. At a given frequency, the nanowire SPP has the largest wavevector confirming thus the high degree of optical confinement achieved in these systems.

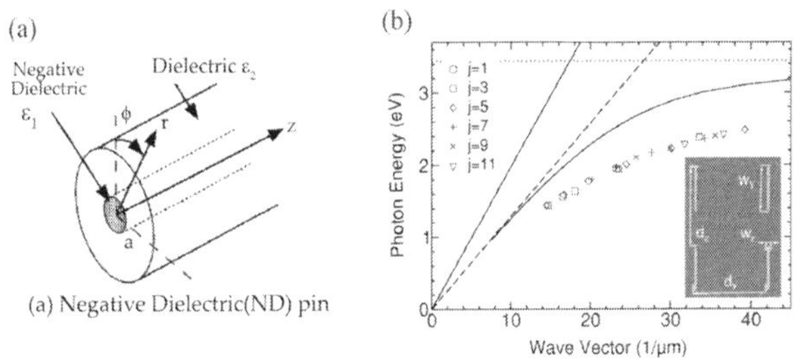


**Figure 13:** (a) Schematic of one dimensional negative dielectric pin model. (b) SPP dispersion relations for different Ag interfaces. The solid line is light line in vacuum. Dashed line is light line in glass. Solid curve is SPP on silver/glass interface. Dotted line is the limiting frequency $\omega_{SP}$ . Symbols are for SPP in Ag nanowires. The wave vector $k = j(\pi/w_y)$, where j is an integer and j values are defined in the legend. $w_y$ is the length of wire. The inset shows the SEM image with the sample geometry..

The SPP dispersion relation is strongly dependent on the size and shape of the nanowires. As the diameter increases, the plasmon resonance frequencies also increases . Nanowires with different cross-sectional shapes have different plasmon resonances. For nanowires with non-regular cross-sections, the electron oscillation can be located along the principle axes or at edges and corners, leading to an additional shape-dependent depolarization and splitting of the surface plasmon resonance into several modes . Figure 14 displays the field amplitude distribution for different cross-sectional nanowire shapes. For the non-cylindrical shapes, the field amplitude distributes mainly at corners.

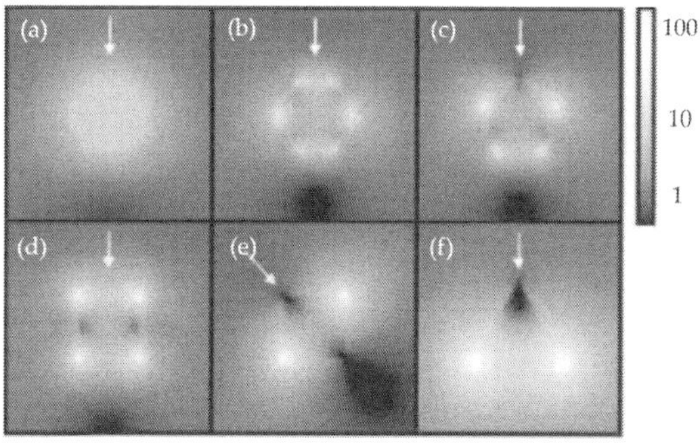


**Figure 14:** Field amplitude distribution for nanowires with different cross-sections at their respective main resonance. (a) circular shape, (b) hexagonal shape, (c) pentagonal shape, (d) and (e) square shape and (f) triangular shape. The diameter of the circular shape is 20 nm. The arrow indicates the propagation direction of the illumination field.

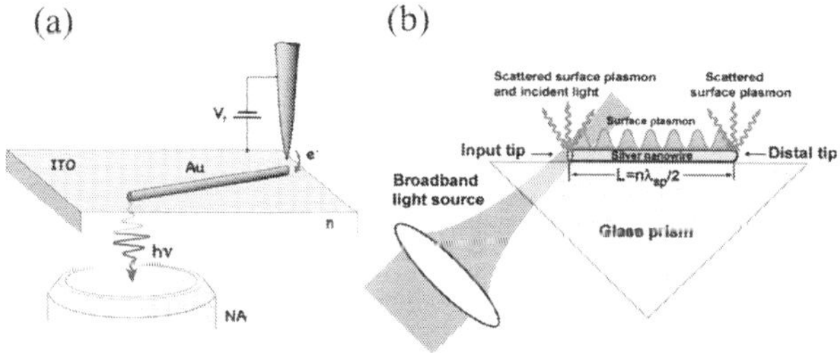


**Figure 15:** Different coupling strategies: (a) Electrical coupling. (b) TIR coupling.

## 6.2.2 Functionalities of nanowires

As a fundamental building block for developing a subwavelength optical circuitry, metallic nanowires have been thoroughly investigated as a basis for active plasmonic devices.

Nanowire-based waveguide with different termination shapes can serve as a polarization rotator. The polarization of light emitted by end-face SPP scattering was affected by the geometry of termination. By changing the end shape of the nanowire through precise chemistry, polarization could either be maintained or rotated. Furthermore, the angular distribution of the scattered SPP indicated that nanowire waveguides could serve as a broadband antenna. Diameter, excitation wavelength, and material dependent unidirectional light emission were found. A bimetallic wire with a sharp Au/Ag heterojunction displays unidirectionality with respect to plasmon propagation across the heterojunction.

Network of nanowires is typically offering a complex control of SPP routing. Crossed nanowire structures composed of a main nanowire and one or several branches were employed for different functionalities. The branched nanowire can serve as a second output of plasmon which functions as a router and multiplexer or serves as an input to launch another plasmon which may function as a modulator (Fig. (Fig. 16(a) and (b)). The realization of different outputs in Fig. 16(b) relies on the control of polarization of the input beam. In Fig. 16(a), the branched nanowire serves as an additional input. If the two input channels have the same polarization, different emission intensities at the network output were obtained due to the phase difference between the two paths. By changing the relative phase or the polarization angle of the two input signals, the signal emission at the output can be modulated. More complicated configuration using a two inputs/two outputs nanowire network was reported very recently. In Fig. 16(b), the network is composed of two inputs $I_1$, $I_2$ and two outputs $O_1$, $O_2$. Images from i to v describe the SPP-coupled fluorescence of quantum dots placed at the two outputs. By changing the polarization and phase of the two input beams, the two output intensities can change in an alternate manner. This strongly depends on the near-field intensity of the branched wire and main wire junction. The physical mechanism underlying these functionalities is the interference of plasmons launched at different positions of this network. The interference of the two plasmons affects the near-field distribution and consequently the output intensity. This kind of multi-input and multi-output nanowire network can be used to realize interferometric logic function.

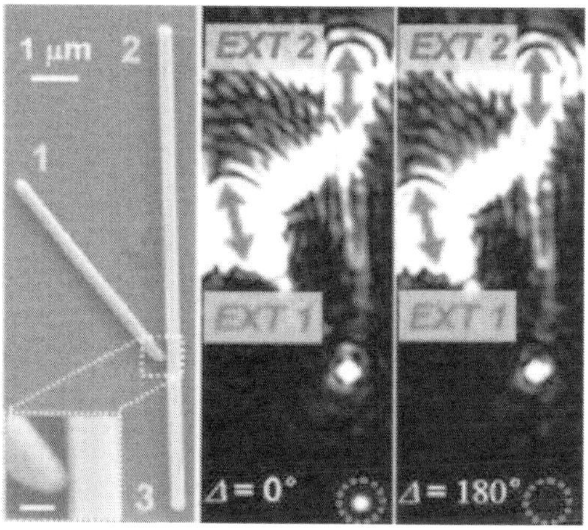


(a)

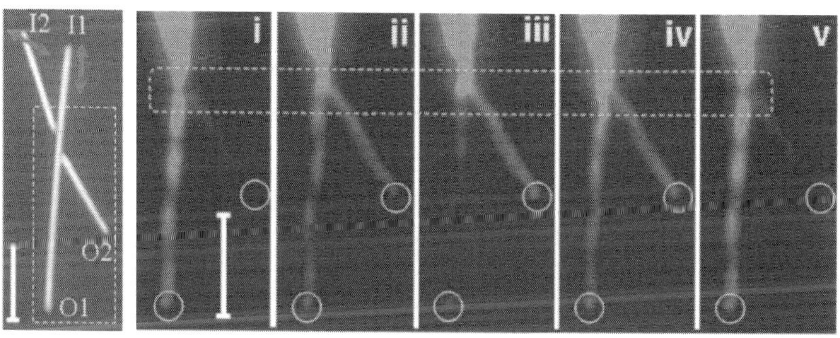


(b)

**Figure 16:** (a) A branched nanowire structure serves as a plasmon modulator. The branch nanowire serves as another input for launching the second plasmon. The arrows show the excitation polarization. (b)Modulation and routing of light in a two inputs/two outputs nanowire network. Images i – v are the quantum dot emission images of one interference cycle.

## 6.3 PLASMONIC DEVICES AND CIRCUITS

### 6.3.1 Plasmonic waveguide circuits

Surface plasmons can propagate as waves over metal surfaces, and because plasmon wavelengths can be much smaller than light at the same frequency, they have potential in highly compact optical devices. The optical properties of these waves on surfaces have been studied and applied to simple optical elements. For example, it was shown that a converging lens for surface plasmons can be created by slits cut into the metal film, which performs an optical Fourier transform analogously to macroscopic lenses, enabling traditional optical computing on-chip . One of the issues with "free" plasmon propagation over surfaces is creating the plasmon in the first place. One solution is to etch a slot through the metal film to the transparent substrate. When the slot is illuminated through the substrate, the electric field penetrates to the top surface, launching plasmon waves that propagate in either direction away from the normal to the slot edge. By including slot reflectors in the surface, it is possible to interfere the plasmon waves enabling devices such as binary encoders , logic discriminators , and multiplexers . Similarly, arrays of holes can be used to launch plasmons and to convert the energy back into freely propagating light .

Plasmon waves can be confined and guided by a thin metal strip. The motivation for such work is based on the idea that light can form high-speed and low-loss interconnects for electronic circuits and the use of plasmons enables very compact devices, an order of magnitude or smaller than the wavelength of light. Similar to light travelling in optical waveguides, the plasmons propagate with one or more different modes on the metal strip, depending on the strip geometry and the electric permittivities of the metal and the surrounding medium. A systematic study of the propagation in these guides, demonstrated the existence of a short-range and a long-range mode. The short-range mode is lossy because there is strong penetration of the plasmon electric field into the metal, creating electrical currents that lose energy by Ohmic resistance. The long-range mode has the plasmon electric field predominantly in the dielectric region about the waveguide, thereby reducing losses. Transmission of optical signals by plasmons in strip waveguides has been demonstrated at telecom wavelengths with data rates of 10 Gbs . There have also been investigations of plasmon propagation in more complicated multi-layer guides, as well as arrays of particles .

As loss is a big problem with plasmon propagation in metals, there have been many studies looking for configurations of metals and dielectrics that minimize the penetration of the electric field into the metal . These include grooves, slots, metal-insulator-metal designs, and so on (see Figure 17 with typical parameters in Table 1). Important factors in plasmon waveguides are the propagation distance, operating wavelength, and mode area.

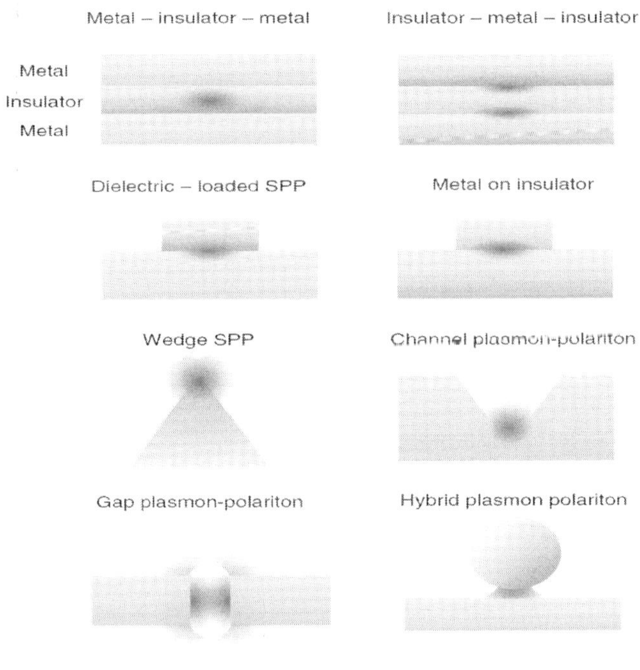


**Figure 17:** Cross sections of different types of plasmon waveguides.

The propagation distance is limited by absorption, which is problematic in complex plasmonic circuits. The mode area describes the cross sectional area of the region containing the optical energy. Large mode areas reduce the ability to create compact devices. For example, subwavelength plasmon-polariton guiding by triangular metal wedges has been demonstrated with propagation lengths ~120 μm and mode widths ~3 μm. Other proposed strategies to mitigate Ohmic losses in plasmon oscillation and propagation include the interaction of plasmons with gain media .

**Table 1:** Comparison of plasmon waveguide optical confinement (mode area), propagation length $L$ and free space wavelength $\lambda$ .

| Waveguide type | Mode width/λ | Mode height/λ | Mode area/ $(\lambda/2)^2$ | $L/\lambda$ | λ (nm) |
|---|---|---|---|---|---|
| GPP | 0.13 | 0.13 | 7% | 13 | 1550 |
| Wedge | ~0.47 | N.A. | N.A. | 2 | 633 |
| MIM | 0.5 | 0.2 | 10% | 5 | 685 |
| IMI | 0.19 | ~0.19 | ~14% | 14 | 1550 |
| V-groove | 0.65 | >0.84 | >200% | ~52 | 1550 |
| DL-SPP | 0.62 | N.A. | N.A. | 10 | 800 |
| DL-SPP | N.A. | N.A. | 3% | 4 | 1550 |
| | 0.04 | 0.04 | 0.6% | 21 | 1550 |
| | 0.14 | 0.04 | 2% | N.A. | 1427 |
| HPP | 0.31 | 0.06 | 7% | 21 | 808 |
| | 0.09 | 0.08 | 3% | 11 | 633 |
| | $0.15_{theory}$ | $0.05_{theory}$ | 3% | 24 | 1310 |
| | | | | | |

Metal-insulator-metal (MIM), sometimes referred to as metal-dielectric-metal (MDM), and gap plasmon polariton (GPP) waveguides have the advantage of simple fabrication in that they can be constructed by cutting a slot in a metal film and filling the slot with a dielectric , . Likewise, dielectric loaded plasmon waveguides only require patterning of a dielectric layer on top of a metal film, which can be done directly using lithographic resist such as PMMA. Guiding in "V"-shaped grooves in metals has been demonstrated along with typical waveguide elements , such as splitters and ring resonators. A theoretical analysis of plasmons in channels showed that increasing wavelength caused the fundamental mode to shift from the bottom of the channel and become more like wedge plasmons, being guided by the upper edges . The plasmons guided by wedges at telecom wavelengths are better confined (smaller mode area) without increasing loss .

The choice of one waveguide geometry over another depends on the application. If long propagation distance is required, then one might choose a V-groove or hybrid plasmon-polariton geometry, or if small mode area is necessary, then an insulator-metal-insulator design may be preferred. However, these waveguides may not be compatible with the overall microfabrication process, in which case some trade-off will be required.

## *Emitters and detectors for integrated plasmonics*

With the development of plasmon waveguide devices, there has been some research into methods of launching plasmons directly on-chip. Most of these works have been proof-of-principle demonstrations of devices with potential for integration in plasmonic circuits, but they are significantly less advanced than corresponding devices used in photonics. In one example, a nanoscale light emitting diode with a subwavelength footprint directed some of its emission into a single-mode slot plasmon waveguide . Additionally, the prospect of stimulated emission of plasmon radiation has been studied , and a device was fabricated and demonstrated . This nanoscale plasmon emitter was constructed from 44-nm-sized nanoparticles with a gold core and a dye-doped silica shell. Surface plasmon oscillations were outcoupled to photonic modes at a wavelength of 531 nm. At the quantum limit, there have been experimental demonstrations of the excitation of surface plasmons by single photon emitters. Importantly, these experiments reveal that the radiative decay of a plasmon excited by a single photon also yields a single photon, even though a surface plasmon is a collective phenomenon consisting of the in-phase oscillations of a large number of electrons.

Active plasmonic devices including those with gain have been reviewed elsewhere . Gain media consist of materials with electronic states that can be optically excited and subsequently de-excited in phase by another light beam, as occurs in solid-state lasers. Plasmon gain requires the gain media to emit in phase with the propagating plasmon. Amplified spontaneous emission from a polymer waveguide film containing laser dye molecules excited by surface plasmons and pumped by another laser has been observed , and there has been a demonstration of plasmonic propagation with net positive gain provided by an optically pumped layer of fluorescent polymer in a dielectric-metal-dielectric waveguide . Incoherent plasmonic gain has been predicted with three-level fluorescent semiconductor nanocrystals in the presence of another semiconductor with negative electric permittivity . Spontaneous emission from nanosized particles and plasmon resonators has been studied theoretically .

Plasmon detectors have also been integrated into devices, such as a Schottky contact device with an asymmetric metal stripe waveguide . Superconducting plasmon detectors were used to detect single plasmon quanta in a quantum interference experiment . Plasmonic components have been used in detectors to alter the responsivity in the mid-infrared as well as in the visible region . In principle, such detectors can be as sensitive as those used in photonics, which is determined largely by the quality of the semiconductor photo-diode fabrication process

## *Linear devices for optical computing*

Linear devices based on waveguides tend to mimic those used in photonics. Linear devices use plasmon wave properties such as constructive and destructive interference to perform addition and subtraction operations, respectively. Moreover, plasmon excitation relies on the correct alignment of the incident electric field with respect to the ridges or grooves that launch the plasmons, as the field is required to induce a surface charge. In other words, such launching methods are sensitive to the incident polarization, which provides a means for polarization sensitivity.

Examples of photonic devices created for plasmons are shown in Figure 18 and include "Y" junctions for combining or separating plasmons,"X" junctions , proximity couplers , directional couplers , Mach-Zehnder interferometers for interfering plasmons, Bragg grating filters , add-drop filters , and ring resonators , that transmit or reflect plasmon waves depending on frequency and switching using phase shifts . A novel plasmon filter was created using two wires of different materials joined together that provided a large electric permittivity mismatch resulting in unidirectional propagation .

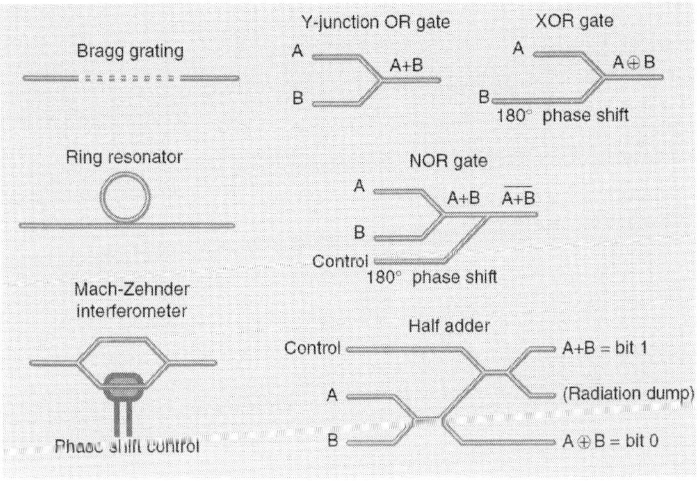


**Figure 18:** Plasmonic waveguide devices for performing optical processing. The logic circuits operate using phase shifts and interference.

Additive logic operations have been proposed and modeled with such circuits or their variants. There have been demonstrations of logic functions such as a NOR gate built from OR and NOT gates ; XNOT, XOR, and NOT gates using an air slot etched in gold on silicon dioxide and an XOR gate ; an OR gate ; a logic comparator ; NOT, AND, OR, XOR ; a half-adder ; demultiplexers ; binary encoders , ; and discriminators . The operations of these gates depend on phase shifts or on frequency (wavelength) differences, which presuppose specific encoding schemes for the digital information. These devices have been used to demonstrate logic operations, but there has been no concerted attempt at fulfilling the requirements of logical optical processes as required in communications networks or optical computing.

More complex arrangements are possible leading to coupling between waveguide modes . Such coupling can be used to create highly compact waveguide interferometers, such as the Mach-Zehnder. As interferometers are very sensitive to phase shifts, they have applications where non-linear effects can be used to modulate the plasmon phase.

### *Non-linear devices for optical computing*

Linear devices have limited use in many photonic circuits, as they cannot demonstrate bistability or static behavior but require a continuous propagation of plasmon waves to operate (such as required for interference). While plasmon-plasmon interactions and plasmon-material interactions are generally linear, the plasmons have electric field strengths one or more orders of magnitude larger than the light wave electric fields that excite them, which makes it easier to induce non-linear behaviors in materials, enhancing modulation .

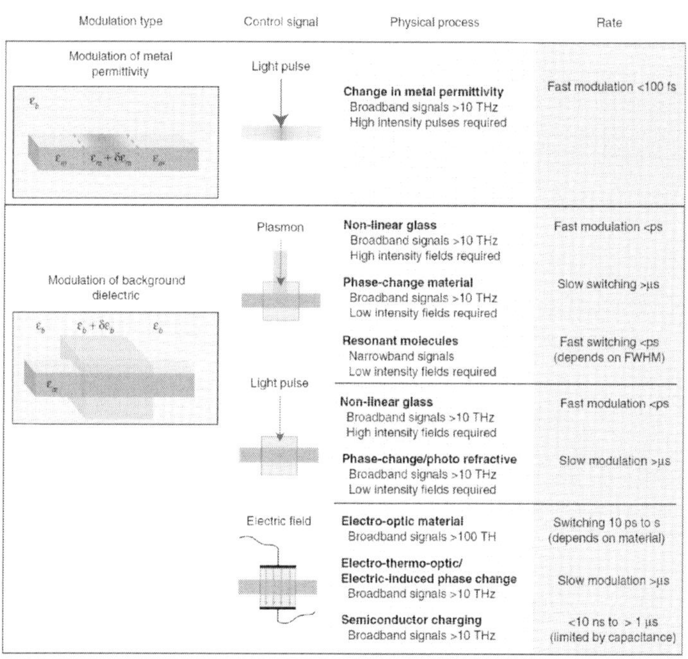


**Figure 19:** Comparison of plasmon modulation methods.

The non-linearities generally change the local electric permittivity that shifts the phase of propagating plasmons. When combined with plasmon interferometers, these can be quite effective modulators. In addition, it is possible to use electrical signals to induce material changes, particularly in semiconductors where the electron density, and therefore the "metallic" properties, can be altered. These are termed "active plasmonic" devices . The methods for non-linear modulation of plasmons are summarized in Figure 19.

There have been many devices proposed and simulated, such as modulation based on an external electrical signal interacting with graphene to create NOR/AND, NAND/OR, XNOR/XOR gates ; third-order non-linear optical materials in which the electric field intensity $|E|^2$ of the plasmon induces refractive index changes, also known as Kerr-nonlinearity intense optical pulses to induce refractive index changes in GaAsInP ; polarization sensitive four-wave mixing non-linearity coupled to a plasmon ring resonator; metal-dielectric cavities filled with non-linear materials to create bistability or switching ; a three-level system showing gain , beam steering by modulating refractive index; electro-active materials to shift refractive index; and change in resonance by changing refractive index or by changing material properties by electron excitation pumped by light , (changes absorption properties within a waveband). One novel device modulated a plasmon wave on a diffraction grating by changing the refractive index of a surface layer of fluid .

It is one thing to propose a device but much more difficult to fabricate and demonstrate one. Plasmonic devices based on non-linear effects can be classified into several different categories: those where the plasmon does the modulation, those where an external light signal does the modulation, and those where an external electrical signal does the modulation. These methods are summarized in Figure 19.

### *Direct plasmon modulation*

Surprisingly, there have been few experiments demonstrating

plasmon-plasmon modulation. As the plasmon is associated with a strong electric field, a propagating plasmon wave can directly induce enough refractive index change in a material to affect the phase of another plasmon. Nevertheless, the Kerr non-linearity of most optical materials, which determines the change in refractive index with optical intensity, is in the order of ~$10^{-17}$m$^2$/W . That is, to change the refractive index by 1% requires light or plasmon intensities around $10^{15}$ W/m$^2$. While this is very large, compressing a 1-W laser beam with a cross section area of 1 mm$^2$ into a region 30×30 nm$^2$ will create such an intensity. A more practical method for modulating one plasmon by another has been demonstrated using semiconductor nanoparticles, which have an electronic excited state near the plasmon frequency. The excitation of an electron into this state by one plasmon changes the absorption and refractive properties experienced by another plasmon at a different frequency, leading to all-plasmon modulation , . An alternative is to use a photochromic dye in which the refractive index is altered by the presence of the plasmon although such processes are slow, taking milliseconds or even seconds to occur.

### *Modulation by external light*

It is well known that light can induce non-linear changes in material properties, which in turn can be used to modify the phase or absorption of propagating electromagnetic waves. In one class of devices, a high intensity light pulse directed onto a metal disturbs the free electron distribution changing the electric permittivity of the metal or an adjacent semiconductor. This is an extremely fast effect, usually induced by femtosecond laser pulses, which perturbs the plasmon . A similar effect was observed recently in indium tin oxide nanorods supporting plasmons in the infrared . There are interferometers in which refractive index modulation by a light pulse on a photo-refractive material induces a phase shift in one beam path changing the interference state and therefore the plasmon intensity. This method can be used with magneto-optical materials . Light pulses can induce phase transitions in materials, such as vanadium dioxide, which are accompanied by large refractive index changes , or light can induce mechanical strain to change the periodicity of a grating, thereby modulating the propagating plasmon . A variant on this method was to use a light pulse to induce scattering centers by the local decomposition of silver oxide .

### *Modulation by an electrical signal*

A plasmon ring resonator was constructed with a dielectric host-matrix doped with an electro-optic material that changes refractive index on application of an electrical signal, modulating the plasmon transmission. Such devices can be slow depending on the electrical response time of the material . Recently, a plasmonic Mach-Zehnder modulator was demonstrated, which used an electro-optic material to modulate the phase. The on-chip modulator was integrated into a silicon waveguide of 10-μm length with a frequency response up to 70 GHz . A novel plasmon memory device was based on memristor technology in which the growth of a metal filament under action of an applied electric field modulates the plasmon propagation in a MIM waveguide structure. It is possible to control the electron density by an electric field by using a semiconductor or electro-optic material, which can be used to modulate plasmons, or by using Tamm-plasmon-polaritons, which are plasmon states formed at the boundary between a metal and a dielectric Bragg mirror . A similar phenomenon is the Faraday effect in which a static magnetic field applied to a magneto-optic material alters the optical properties, which can be coupled to plasmonic devices .

Variants of this method use electrical signals to induce temperature changes that modify the refractive index of a thermo-optic dielectric, again by using interferometry to alter the plasmon intensity or by using electrochemical switching of material properties , .

It is clear that the most promising methods for plasmon modulation use electrical effects. These are easily integrated into existing electronic circuit designs and can show superior performance .

## 6.3.2 Hybrid Plasmonic-Photonic Nanowire Devices

A hybrid plasmonic waveguide is an optical waveguide that achieves strong light confinement by coupling the light guided by a dielectric waveguide and a plasmonic waveguide. It is formed by separating a medium of high refractive index (usually silicon) from a metal surface (usually gold or silver) by a small gap.

**Figure 20**: Cross section of hybrid plasmonic waveguide. Power propagates in the z direction.

### *History*

Dielectric waveguides use total internal reflection to confine light in a high index region. They can guide light over a long distance with very low loss, but their light confinement ability is limited by diffraction. Plasmonic waveguides, on the other hand, use surface plasmon to confine light near a metal surface. The light confinement ability of plasmonic waveguides is not limited by diffraction, and, as a result, they can confine light to very small volumes. However, these guides suffer significant propagation loss because of the presence of metal as part of the guiding structure. The aim of designing the hybrid plasmonic waveguide was to combine these two different wave guiding schemes and achieve high light confinement without suffering large loss. Many different variations of this structure have been proposed. Many other types of hybrid plasmonic waveguides have been proposed since then to improve light confinement ability or to reduce fabrication complexity.

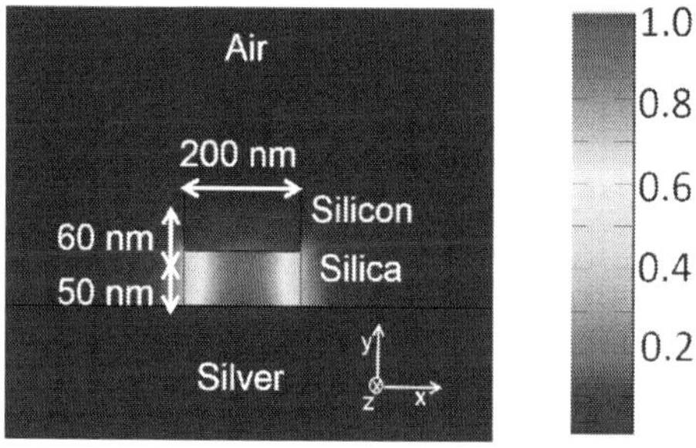


**Figure 21:** Guided power density in a hybrid plasmonic waveguide. Light propagates in the z-direction

## *Principle of operation*

The operation of the hybrid plasmonic waveguides can be explained using the concept of mode coupling. The most commonly used hybrid plasmonic waveguide consists of a silicon nanowire placed very near a metal surface and separated by a low index region. The silicon waveguide supports dielectric waveguide mode, which is mostly confined in silicon. The metal surface supports surface plasmon, which is confined near the metal surface. When these two structures are brought close to each other, the dielectric waveguide mode supported by the silicon nanowire couples to the surface plasmon mode supported by the metal surface. As a result of this mode coupling, light becomes highly confined in the region between the metal and the high index region (silicon nanowire).

## *Applications*

Hybrid plasmonic waveguide provides large confinement of light at a lower loss compared to many previously reported plasmonic waveguides. It is also compatible with silicon photonics technology, and can be integrated with silicon waveguides on the same chip. Similar to a slot-waveguide, it can also confine light in the low index medium. Combination of these attractive features has stimulated worldwide research activity on the application of this new guiding scheme. Some notable examples of such applications are compact lasers, electro optic modulators, biosensors, polarization control devices, and thermo-optic switches.

## REFERENCES


1. Akimov, Yu A; Chu, H S (2012). "Plasmon–plasmon interaction: Controlling light at nanoscale". Nanotechnology. 23 (44): 444004. doi:10.1088/0957-4484/23/44/444004. PMID 23080049.
2. Barnes, William L.; Dereux, Alain; Ebbesen, Thomas W. (2003). "Surface plasmon subwavelength optics" (PDF). Nature. 424 (6950): 824–30. Bibcode:2003Natur.424..824B. doi:10.1038/nature01937. PMID 12917696. Archived from the original (PDF) on 2011-08-11. Free PDF download.
3. Barnes, William L.; Dereux, Alain; Ebbesen, Thomas W. (2003). "Surface plasmon subwavelength optics". Nature. 424 (6950): 824–30. Bibcode:2003Natur.424..824B. doi:10.1038/nature01937. PMID 12917696. S2CID 116017.
4. Bashevoy, M.V.; Jonsson, F.; Krasavin, A.V.; Zheludev, N.I.; Chen Y.; Stockman M.I. (2006). "Generation of traveling surface plasmon waves by free-electron impact". Nano Letters. 6 (6): 1113–5. arXiv:physics/0604227. doi:10.1021/nl060941v. PMID 16771563.
5. Cottam, Michael G. (1989). Introduction to Surface and Superlattice Excitations. New York: Cambridge University Press. ISBN 978-0750305884.
6. Homola, Jirí (2006). Surface Plasmon Resonance Based Sensors. Springer Series on Chemical Sensors and Biosensors, 4. Berlin: Springer-Verlag. ISBN 978-3-540-33918-2.
7. Huidobro, Paloma A.; Nesterov, Maxim L.; Martín-Moreno, Luis; García-Vidal, Francisco J. (2010). "Transformation Optics for Plasmonics" (PDF). Nano Letters. 10 (6): 1985–90. arXiv:1003.1154. Bibcode:2010NanoL..10.1985H. doi:10.1021/nl100800c. hdl:10044/1/42407. PMID 20465271. Free PDF download for these peer reviewed articles.
8. Minh Hiep, Ha; Endo, Tatsuro; Kerman, Kagan; Chikae, Miyuki; Kim, Do-Kyun; Yamamura, Shohei; Takamura, Yuzuru; Tamiya, Eiichi (2007). "A localized surface plasmon resonance based immunosensor for the detection of casein

in milk". Science and Technology of Advanced Materials. 8 (4): 331. Bibcode:2007STAdM...8..331M. doi:10.1016/j.stam.2006.12.010.

9. NIST researchers, Nanofabrication Research Group (2009-08-20). "Three-Dimensional Plasmonic Metamaterials". National Institute of Science and Technology. Retrieved 2011-02-15.
10. NIST researchers, Nanofabrication Research Group (2010-02-11). "Opto-mechanical Devices for Measuring Nanoplasmonic Metamaterials". National Institute of Science and Technology. Retrieved 2011-02-15.
11. NIST researchers, Nanofabrication Research Group. "Nanoplasmonics" (Online). National Institute of Science and Technology. Retrieved 2011-02-15.
12. Ozbay, E. (2006). "Plasmonics: Merging Photonics and Electronics at Nanoscale Dimensions". Science. 311 (5758): 189–93. Bibcode:2006Sci...311..189O. doi:10.1126/science.1114849. hdl:11693/38263. PMID 16410515. S2CID 2107839.
13. Pan, Bai Cao; Liao, Zhen; Zhao, Jie; Cui, Tie Jun (2014). "Controlling rejections of spoof surface plasmon polaritons using metamaterial particles". Optics Express. 22 (11): 13940–13950. doi:10.1364/OE.22.013940. PMID 24921585.
14. PDF from arxiv.org – Transformation Optics for Plasmonics. 15 pages.
15. Pendry, J. B.; Martín-Moreno, L.; Garcia-Vidal, F. J. (6 August 2004). "Mimicking Surface Plasmons with Structured Surfaces". Science. 305 (5685): 847–848. Bibcode:2004Sci...305..847P. doi:10.1126/science.1098999. PMID 15247438.
16. Pitarke, J M; Silkin, V M; Chulkov, E V; Echenique, P M (2007). "Theory of surface plasmons and surface-plasmon polaritons" (PDF). Reports on Progress in Physics. 70 (1): 1. arXiv:cond-mat/0611257. Bibcode:2007RPPh...70....1P. doi:10.1088/0034-4885/70/1/R01. Free PDF download.
17. Pochi Yeh (3 March 2005). Optical Waves in Layered Media. Wiley. ISBN 978-0-471-73192-4.
18. S.Zeng; Baillargeat, Dominique; Ho, Ho-Pui; Yong, Ken-Tye; et al. (2014). "Nanomaterials enhanced surface plasmon

resonance for biological and chemical sensing applications" (PDF). Chemical Society Reviews. 43 (10): 3426–3452. doi:10.1039/C3CS60479A. PMID 24549396.

19. This article incorporates public domain material from the National Institute of Standards and Technology document: "Three-Dimensional Plasmonic Metamaterials".
20. V. K. Valev (2012). "Characterization of Nanostructured Plasmonic Surfaces with Second Harmonic Generation". Langmuir. 28 (44): 15454–15471. doi:10.1021/la302485c. PMID 22889193.
21. Vo-Dinh, Tuan (2017). "Chapter 13 - Biomolecule Sensing Using Surface Plasmon Resonance". Nanotechnology in Biology and Medicine Methods, Devices, and Applications, Second Edition. United States: CRC Press. pp. 259–288. ISBN 978-1439893784.
22. Wenshan Cai; Justin S. White & Mark L. Brongersma (2009). "Compact, High-Speed and Power-Efficient Electrooptic Plasmonic Modulators". Nano Letters. 9 (12): 4403–11. Bibcode:2009NanoL...9.4403C. doi:10.1021/nl902701b. PMID 19827771.
23. Yarris, Lynn (2009-08-20). "GRIN Plasmonics..." (Online news release). U.S. Department of Energy National Laboratory Operated by the University of California. Retrieved 2011-02-15.
24. Zeng, Shuwen; Yu, Xia; Law, Wing-Cheung; Zhang, Yating; Hu, Rui; Dinh, Xuan-Quyen; Ho, Ho-Pui; Yong, Ken-Tye (2013). "Size dependence of Au NP-enhanced surface plasmon resonance based on differential phase measurement". Sensors and Actuators B: Chemical. 176: 1128–1133. doi:10.1016/j.snb.2012.09.073.

CHAPTER 7

# GAIN-ASSISTED SURFACE PLASMON RESONANCES AND PROPAGATION

## INTRODUCTION

Surface plasmons (SPs) are coherent delocalized electron oscillations that exist at the interface between any two materials where the real part of the dielectric function changes sign across the interface (e.g. a metal-dielectric interface, such as a metal sheet in air). SPs have lower energy than bulk (or volume) plasmons which quantise the longitudinal electron oscillations about positive ion cores within the bulk of an electron gas (or plasma). The capability to confine light into a sub wavelength spatial dimension through plasmonic structures makes manipulating light at nanoscale possible. Various novel phenomena such as surface-enhanced fluorescence and Raman scattering, SP resonance–induced optical force enhancement, and extraordinary transmission through metallic nanoarrays have been discovered due to the excitation of SPs.

Because of their ability to confine light into a sub wavelength spatial dimension, as well as their fast propagation speed (up to 100 THz), nanoplasmonic structures have been regarded as one of the most promising candidates for next-generation information communication technology. They can either be used as building blocks to realize all-optical integrated photonic chip or be integrated with the silicon electronics in a fully compatible manner to solve some critical issues in current computer architectures such as highspeed interconnect at both intra- and interchip levels. With the surge of on plasmonic circuits in recent years, both passive and active plasmonic components have been demonstrated, such as plasmonic waveguides, wavelength-dependent plasmonic routers, plasmon-based modulators, single-plasmon sources, and even plasmonic nanowire networks to perform logic algorithms.

However, metallic loss that accompanies all plasmonic devices has hindered further integration of plasmonic-based components into practical circuits. In the most recent years, researchers have tried to partially replace or completely replace the normal dielectric part of the plasmonic devices with gain materials in order to compensate for the Ohmic loss caused by the metallic components within the devices. Great successes have been achieved in the areas of amplification of long-range surface plasmon polaritons (LRSPPs) using fluorescent materials, stimulated emission from metal-dielectric core-shell nanoparticles, and stimulated emission from hybrid SP modes at the metal–insulator–semiconductor interfaces. Our group has demonstrated that hybrid plasmonic signals of different polarizations can be amplified on the same device through the pump-probe ultrafast technique.

## 7.1 AMPLIFICATION OF LONG-RANGE SURFACE PLASMON POLARITONS

To minimize the metallic loss that is associated with the SPP modes, some researchers sacrifice the strong mode confinement offered by short-range SPPs and trade it for low propagation loss. The result of this compromise is a mode called LRSPP. To obtain LRSPPs, a sufficiently thin metal film needs to be buried in the dielectric. The mathematical description of the LRSPP can be relatively simple. For any SPP mode existing at the interface of two infinite half media, metal and dielectric, its propagation constant β (parallel to the interface) is written as $\beta=\frac{\omega}{c}\sqrt{\frac{\varepsilon_m\varepsilon_d}{\varepsilon_m+\varepsilon_d}}$, where ω and c are the frequency and speed of electromagnetic waves in vacuum, respectively, and $\varepsilon_d$ and $\varepsilon_m$ are the permittivities of the dielectric and metal, respectively. When a thin metal film is embedded in dielectric, the two identical SPP modes at the upper and lower interfaces will couple together and form symmetrical and antisymmetrical modes, whose propagation constants parallel to the interface are implicitly described in the dispersion relation:

$$\tanh(S_m t) = -\frac{2\varepsilon_d S_d \varepsilon_m S_m}{\varepsilon_d^2 S_m^2 + \varepsilon_m^2 S_d^2},\ S_d = \sqrt{\beta^2 - \varepsilon_d k_0^2},\ S_m = \sqrt{\beta^2 - \varepsilon_m k_0^2}, \quad (1)$$

where t is the thickness of the metal film and $k_0 = \omega/c$ is the wave number of light in vacuum. For the symmetric mode, most of the mode profile is within the cladding dielectric layer and is only weakly bound to the metal film. With the decrease in the metal thickness, the part of the mode field within the metal also becomes progressively smaller, resulting in drastically decreased mode absorption and propagation loss. In general, the propagation length of LRSPPs can increase more than an order of magnitude compared to that of a short-range SPP mode.

Since the propagation loss for the LRSPP is usually under 100 $cm^{-1}$, which is feasible with the highly fluorescent dye molecules or conjugated polymers, the experimental efforts are mostly focused on the construction of symmetric (refractive-index-matched) active dielectric layers at the top and bottom interfaces of the metal stripe. Gather adapted the dielectric-metal-dielectric configuration for the LRSPP waveguide, in which the fluorescent conjugated polymers MDMO-poly (phenylene vinylene) blended with the nonfluorescent poly (spirofluorene) (PSF) were used as the gain material. In this structure, the metal layer is 4 nm thick Au. The authors replaced part of the PSF with a low-refractive-index compound to index-match the gain material with that of the dielectric cladding layers. Using this method, the authors could greatly reduce the propagation loss of the LRSPP mode, as shown in Figure 1. A maximum gain of ~59 $cm^{-1}$ at wavelength 600 nm is realized in the experiment.

Almost at the same time, De Leon reported another method to index-match the refractive index of the gain material. The authors fabricated an Au stripe of 20 nm thick on top of the $SiO_2$ and covered the structure with a gain layer of ~100 μm thick IR140 dye solution, as shown in Figure 2. The solvent they used was a mixture of 30.4% ethylene glycol and 69.6% dimethyl sulphoxide. At room temperature, the dye solution was index-matched to $SiO_2$ at the emission wavelength, forming a symmetric LRSPP structure. The authors claimed a gain coefficient of 8.85 dB $mm^{-1}$ in their structure.

These experiments demonstrated that the gain compensation for the propagation loss of LRSPPs can be achieved by highly fluorescent dye molecules or semiconducting conjugated polymers as long as a proper index-matching technique is implemented. The gain coefficients shown in these 100 $cm^{-1}$, still far what is needed for short-range SPPs.

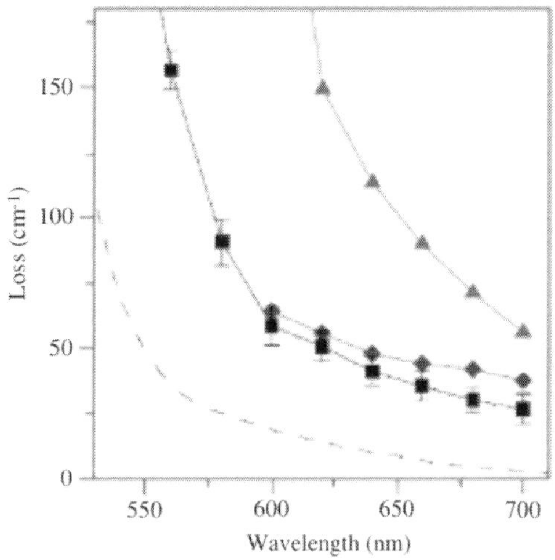


**Figure 1**. Measurement and optimization of the propagation loss in 4 nm thick gold LRSPP waveguides.

Propagation loss (as derived from the exponential decay of scattered light intensity) for different plasmonic waveguides as a function of the wavelength for a structure without a gain material (squares), with a gain layer that was partly index-matched (triangles, nominal Dn=0.03) and fully index-matched (diamonds). Error bars represent uncertainty of the fit to the exponential decay and are representative for all data sets. The dashed line represents a theoretical loss estimate (neglecting scattering) for the structure without a gain material.

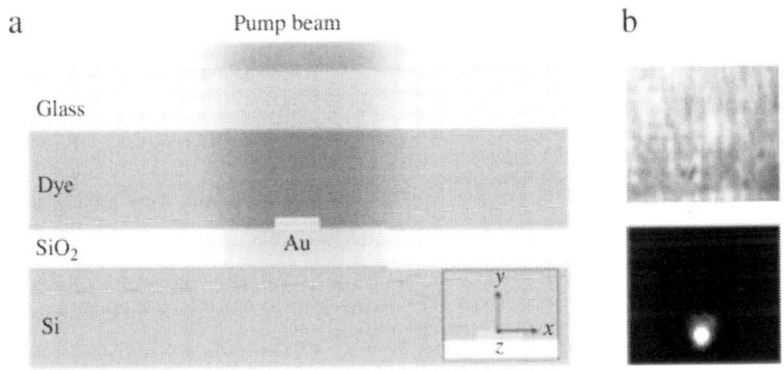


**Figure 2**. Guiding structure.

(a) Cross-sectional view of the active structure (not to scale). The gain medium is in the form of a laser dye in a solution. Inset: coordinate system with the +z direction coming out of the page. (b) x-Polarized (top) and y-polarized (bottom) light collected from the output facet of the structure while being pumped by x-polarized light.

## 7.2 STIMULATED EMISSION FROM TOCALIZED SURFACE PLASMON RESONANCE WITH A GAIN MATERIAL

Due to its potential application in plasmon-based molecular sensors, the enhancement of localized SP resonance with a gain material has gained increasing attention in the plasmon community. The estimation for the critical gain coefficient has been worked out in a few theoretical papers, focusing on different geometries of the nanoparticles. They all start from the calculation of the polarizability of a nanoparticle in an applied electric field. One approximation that people usually make to simplify the calculation is the electrostatic approach, which is valid when the size of the metallic nanoparticle is much smaller than the wavelength of the radiation light. In this case, the electric potential $\Phi$ becomes the solution of Laplace equation $\Delta^2\Phi = 0$, from which we can find the electric field distribution as $E = \Delta\Phi$. A more general form of the polarizability $\alpha$ of an isolated nanoparticle in the electric field is given by$\alpha = (4\pi)^{-1} (\varepsilon_m - \varepsilon_d)/[\varepsilon_d + p(\varepsilon_m - \varepsilon_d)]$, where $\varepsilon_m$ and $\varepsilon_d$ are respectively the complex permittivities of the metal and surrounding dielectric and p is the depolarization factor associated with the shape of the nanoparticle. In the case of a spherical nanoparticle, p = 1/3, and we get $\alpha \propto (\varepsilon_m - \varepsilon_d)/(2_{\varepsilon d} + \varepsilon_m)$.. We can clearly that the resonance condition of the polarizability is satisfied when the value of the denominator $|\varepsilon_d + p(\varepsilon_m - \varepsilon_d)|$ is a minimum. In the case of small or slowly varying $Im(\varepsilon_m)$, the minimum is achieved when $Re(\varepsilon_m) = -(1 - p)\, Re(\varepsilon_d)/p$. The resonance is limited by the nonvanishing term $Im(\varepsilon_m)$. However, when the passive dielectric material is replaced by a gain medium, the $Im(\varepsilon_d)$ is no longer zero but a negative number while being pumped. It can lead to the complete cancellation of the denominator term when $Im(\varepsilon_m) = -(1 - p)Im(\varepsilon_d)/p$ is satisfied, and the resonance will be strongly enhanced and only be limited by the gain saturation. The critical gain coefficient needed to make the propagation loss zero is found to be $\gamma = (2\pi/\lambda_0)Im(\varepsilon_d)/(Re[\varepsilon_d])1/2$, where $\lambda_0$ is the vacuum wavelength at the resonance frequency. If we use the Drude formula to approximate the dielectric function of the metal $\varepsilon_m = 1 - w_p^{\ 2}/[\omega(\omega + i\Gamma)]$, the gain coefficient $\gamma = (2\pi/n\lambda_0)(\Gamma/\omega_p)[p/(1 - p)][1 + n^2(1 - p)/p]^{3/2}$. For spherical nanoparticles, $\gamma = (\pi/n\lambda_0)\,(\Gamma/\omega_p)(1 + 2n^2)^{3/2}$. If we use $\lambda_0 = 630$ nm and n = 1.33 and $\Gamma$ and $\omega_p$ are typical values for gold and silver, we can estimate $\gamma = \sim 10^3\ cm^{-1}$.

This critical gain value is achievable with highly fluorescent materials with a stimulated cross section of approximately 5 ×

$10^{-16}$ $cm^2$, such as some conjugated polymers, dye molecules, or quantum dots (QDs). If we assume each excited molecule adsorbed on a nanoparticle a few nanometers in diameter, it gives the gain coefficient at the order of magnitude needed to compensate for the metallic loss at resonance.

Works regarding the increase of fluorescence intensity for dye molecules adsorbed on islands and films of nanoparticles whose plasmon resonance matches with the absorption band of the molecules were reported The increase of absorption of molecules and QDs and the enhancement of their fluorescence intensity and modification to their emission spectra due to the presence of metallic nanostructures have been investigated intensively in recent years. Studies showed that if metallic nanoparticles were dispersed into an optical cavity, the pump power that is used to reach the threshold for lasing can be greatly decreased.

The authors discussed in detail the influence of Ag nanoparticle aggregates on various properties of a rhodamine 6G (R6G) solution. They studied the absorption, Rayleigh scattering, spontaneous emission, and stimulated emission from the R6G and Ag nanoparticle mixture separately. Their results showed that the increased scattering in the solution is due to an enhancement in the Rayleigh scattering cross section of the metallic particles in the presence of gain media. This scattering intensity can increase or decrease with different pump power values, which is explained with a theoretical model developed. The increased spontaneous emission intensity from R6G is due to the enhanced absorption caused by the field enhancement associated with the SP resonances of Ag nanoparticles. The enhanced stimulated emission, which has the same nature as the enhanced absorption, is observed in both pump-probe and laser experiments. The most interesting result of their experiment is that in the presence of Ag aggregates, the value of amplification is not limited by the saturation-level characteristic of a pure dye solution but grows to a higher magnitude.

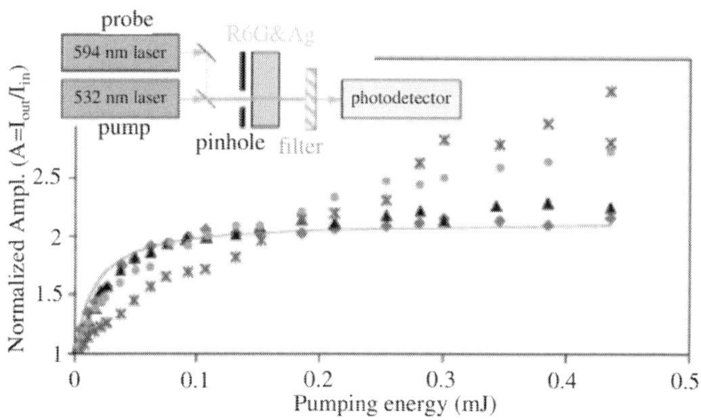


**Figure 3**. Amplification $A = I_{out}/I_{in}$ (at $\lambda = 594$ nm) as a function of the 532 nm pumping energy (after a 0.5 mm pinhole) measured in a series of R6G dye–Ag aggregate solutions in a 10 mm cuvette. Solid line: calculation corresponding to a pure dye, $1.25 \times 10^{-5}$ M. Characters: experiment. Diamonds: pure dye, $1.25 \times 10^{-5}$ M; triangles: dye, $1.25 \times 10^{-5}$ M, Ag aggregate, $3.6 \times 10^{11}$ particles/cm3; circles: dye, 9.0 × 10–6 M, Ag aggregate, $2.5 \times 10^{12}$ particles/cm3; crosses: dye, $5.1 \times 10^{-6}$ M, Ag aggregate, $5.2 \times 10^{12}$ particles/cm3. All amplification signals are normalized to the transmission in the corresponding not-pumped media. Inset: schematic of the experiment.

The concept of surface plasmon amplification by stimulated emission of radiation (SPASER) in a nanosystem was first explicitly predicted. They stated that the radiation of SPASER is mostly composed of SPs, which undergo stimulated emission as photons, only that they can be confined to nanoscale. In SPASER, the active medium undergoes population inversion upon strong pump and transfers its excited energy to a resonant nanosystem via nonradiative transitions. The energy transfer from an excited molecule to a resonant SP oscillation in a metallic nanostructure can also be explained by Förster dipole– dipole interaction, which can also give rise to the main features of SPASER.

SPASER was first demonstrated experimentally. The authors synthesized high-brightness luminescentcore-shell nanoparticles, which are composed of a gold core, providing the plasmon modes, surrounded by a silica shell containing the Oregon Green 488 (OG-488) dye molecules, as the gain medium.

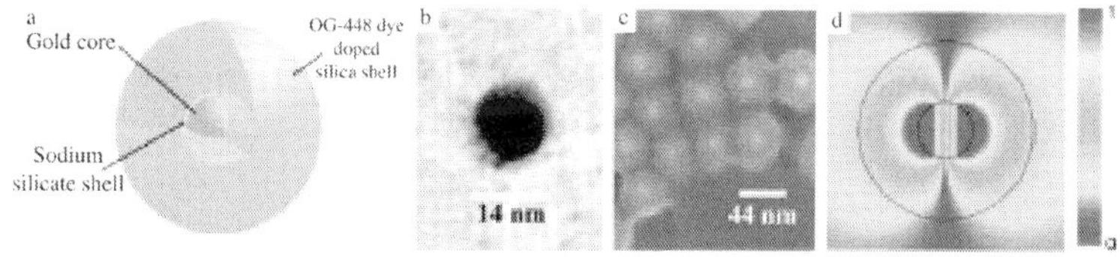


**Figure 4**. (a) Diagram of the hybrid nanoparticle architecture (not to scale), indicating dye molecules throughout the silica shell. (b) Transmission electron microscope image of Au core. (c) Scanning electron microscope image of Au/ silica/dye core-shell nanoparticles. (d) SPASER mode (in a false color), with $\lambda = 525$ nm and $Q = 14.8$; the inner and outer circles represent the 14 nm core and the 44 nm shell, respectively. The field strength color scheme is shown on the right.

The diameter of the Au core is ~14 nm and that of the silica shell ~15 nm. There are roughly $2.7 \times 10^3$ molecules per nanoparticle. These nanoparticles were suspended in a water solution after cleaning, with a concentration equal to 3 ¥ $10^{11}$ $cm^{-3}$. In this configuration, the extinction spectrum of the nanoparticle in the water solution overlaps with both the excitation and emission bands of the dye molecules incorporated in the silica shell. At a stronger pump intensity, the emission spectrum of the solution undergoes a characteristic peak narrowing, corresponding to the commencement of stimulated emission. The authors then went on showing that the full-width at half-maximum (FWHM) of the stimulated emission peak and the emission intensity changed with the pump energy following the same trend as a regular laser. To ensure that this observation is not caused by random lasing within a solution or a feedback provided by the cuvette walls, where the solution was contained, the authors diluted the solution by 100-fold and still observed the same spectral line without diminishing the ratio of stimulated emission intensity to spontaneous emission background. The authors then concluded that the emission spectra are characteristic of each nanoparticle.

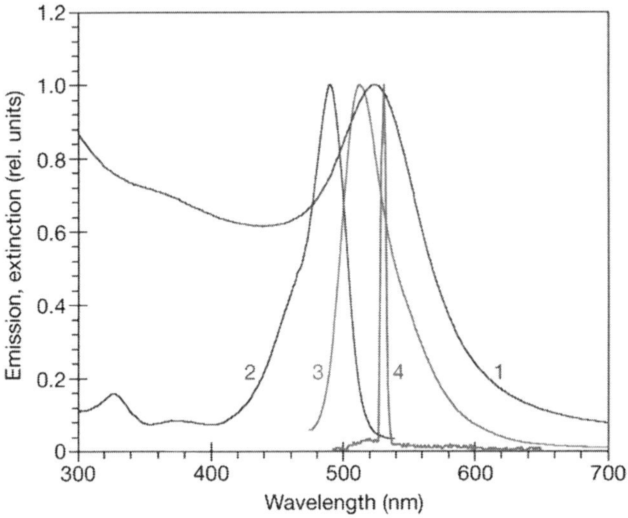


**Figure 5**. Normalized extinction (1), excitation (2), spontaneous emission (3), and stimulated emission (4) spectra of Au/silica/dye nanoparticles. The peak extinction cross section of the nanoparticles is $1.1 \times 10^{-12}$ $cm^2$. The emission and excitation spectra were measured in a spectrofluorometer at low fluency.

The authors then estimated the number of excited molecules per nanoparticles that is needed to compensate for the loss in the localized SP resonance, which is about $2\times 10^3$ molecules per nanoparticle, following a similar method as presented at the beginning of this section. This number is less than the available number of molecules that are incorporated in the silica shell. The pump energy that the authors used to achieve lasing is that needed to saturate all OG-488 molecules. Therefore, it is assumed that all molecules within each nanoparticle are in their excited states. The authors then finally concluded that they have demonstrated a SPASER "nanolaser that is realized by each individual nanoparticle, making it the smallest reported in the only one to date operating in the visible range."

This result has quite a significant impact on the plasmon community. Its demonstration means that it is possible to have a coherent light source whose size is comparable with the typical size of a function unit in the current complementary metal-oxide-semiconductor (CMOS) industry and promises a bright future for plasmon-based integrated photonics.

Finishing up this section, there is one more experiment we would like to mention. Walther demonstrated an electrically pumped microcavity terahertz laser based on an LC circuit resonant oscillation with a mode volume at a deep sub wavelength scale. The Au metal cap on top of the active medium (quantum cascade laser) makes it possible to confine the electric field to the gain region. The simulation of electric field distribution shows a localized resonance at the active medium. They estimated that the spontaneous emission life time of the gain medium is enhanced 17-fold. The quality factor they obtained in the experiment is from 11 to 21, a value that is limited by the metallic loss. On the basis of the same working principles, they speculated that similar lasers can be realized at higher frequencies and even into the near-infrared range.

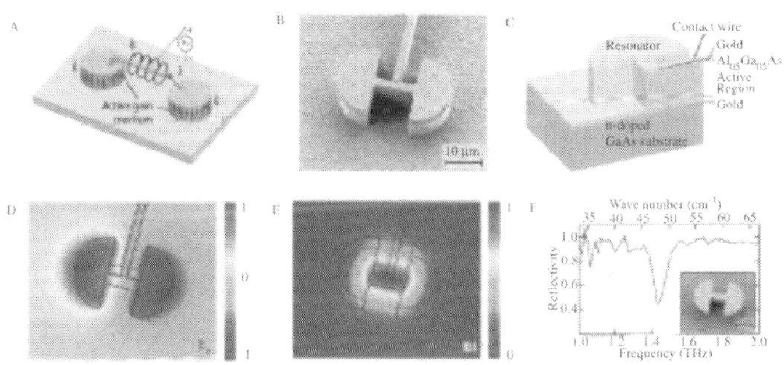


**Figure 6**. (A) Schematic of the LC laser.

J is the alternating current in the resonator, B is the induced magnetic field, and E is the electric field. The active gain medium is biased by the voltage source $V_{DC}$. (B) Scanning electron micrograph of the LC laser device. (C) Schematic cross section through the device along the symmetry axis. The red layer is undoped $A_{l0.5}Ga_{0.5}As$ and prevents current injection into the active region the bonding pad. (D, E) Finite element simulations of the electromagnetic field in the resonator showing the dominating electric field component $E_z$

and the norm of the magnetic field |B|. (F) Measured reflectivity at 10 K of an array of 400 identical LC resonators, shown in the inset and designed for a frequency of 1.45 THz, without a gain medium and without an electrical connection.

## 7.3. GAIN-ASSISTED HYBRID SURFACE PLASMON PROPAGATION: LASING AND AMPLIFICATION

The propagation of surface plasmon polaritons on metallic waveguides adjacent to a gain medium is considered. It is shown that the presence of the gain medium can compensate for the absorption losses in the metal. The conditions for existence of a surface plasmon polariton and its lossless propagation and wavefront behavior are derived analytically for a single infinite metal-gain boundary.

### 7.3.1. Hybrid Surface Plasmon Lasers

In parallel with the works presented in last section, people also tried to increase the propagation length of propagating SPPs, which may serve as optical signal carriers with a sub wavelength mode confinement, with gain materials. In this section, we will discuss the lasing and amplification of a loss propagating SPP mode.

For localized SP resonance, the compensation of loss by the gain material is shown as the enhanced polarizability and narrowing of resonance peak line width. For propagating SPPs, people first studied the enhanced propagation length upon the implementation of gain media with various waveguide configurations. For nanoparticle chains embedded in a gain medium, Citrin showed that the increase in interparticle coupling strength caused by the active material can lead to an extended propagation distance. Avrutsky theoretically investigated SPPs at the metal–dielectric interface with a strong optical gain. Interestingly, he predicted that at the resonance condition, the effective refractive index of the SPPs can go to infinity, which is only limited by the gain saturation, and results in strong localization of the surface wave at the interface. This effect can be simply demonstrated with the dispersion relation of SPPs existing at the metal–dielectric interface $n_{eff} = \sqrt{\frac{\varepsilon_m \varepsilon_d}{\varepsilon_m + \varepsilon_d}}$. As we discussed in the localized SP resonance case, if the dielectric is carefully chosen so that $Re(\varepsilon_d) = Re(\varepsilon_m)$, the resonance condition will be fulfilled. With optical gain in the same dielectric, the imaginary part of the two materials can also be cancelled, meaning $Im(\varepsilon_d) = Im(\varepsilon_m)$. Then $n_{eff}$ becomes infinitely large, making the SP mode extremely localized at the interface. However, this phenomenon is yet to be demonstrated by the experiment. For materials not satisfying the resonance condition, the gain coefficient is written as $\gamma = k_0 Im(\varepsilon_d)/\sqrt{Re(\varepsilon_d)}$, where $k_0$ is the wave number of light in vacuum. For semiconductor materials with n = 3.4, $\gamma$ is expected to be ~4590 $cm^{-1}$ for lossless propagation at the semiconductor–Ag interface with a wavelength of 650 nm, even higher than that of a localized SP resonance case and just falling within the limits that can be offered by semiconductor quantum wells and QDs. However, for lower-refractive-index materials such as doped glass or conjugated polymers, this requirement can be lowered to<1000 $cm^{-1}$.

To maintain the deep sub wavelength confinement offered by the short-range SPPs and be able to further decrease the propagation loss in the SPP mode, Oulton proposed theoretically a hybrid plasmonic mode, which exists at the metal–insulator–semiconductor interfaces. This hybrid plasmonic waveguide was then demonstrated showing mode areas as small as $\lambda^2/400$. The authors also demonstrated the full compensation of propagation loss of this hybrid mode by the realization of SPASER at 489 nm at a low temperature of<10K.

To illustrate the unique properties of a hybrid plasmonic mode, the authors compared the plasmon lasers with conventional photonic lasers, which consist of CdS nanowires on a quartz substrate, as shown in Figure 7. As there is literally no cutoff frequency for the hybrid plasmonic waveguide mode, it was shown clearly that for the plasmon lasers, the lasing threshold is only weakly dependent on the diameter of the nanowire, while for photonic lasers the lasing threshold is strongly dependent on the diameter of the nanowire, owing to the reduction in the total gain volume for smaller nanowires.

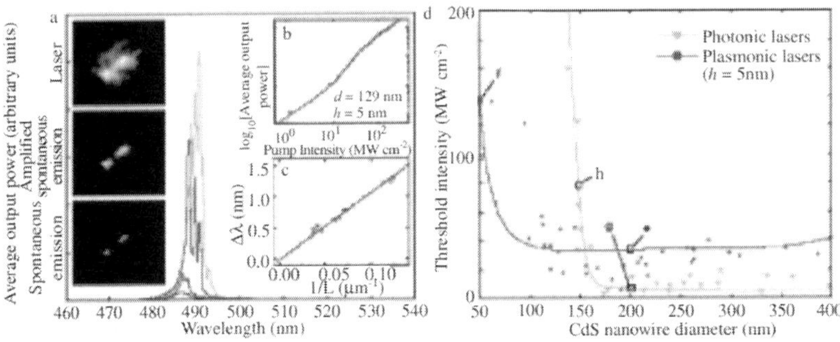


**Figure 7.** Laser oscillation and threshold characteristics of plasmonic and photonic lasers.

(a) Laser oscillation of a plasmonic laser, d = 129 nm and h = 55 nm (longitudinal modes). The four spectra for different peak pump intensities exemplify the transition from spontaneous emission (21.25 MW $cm^{-2}$) via amplified spontaneous emission (32.50 MW $cm^{-2}$) to full laser oscillation (76.25 MW cm–2 and 131.25 MW $cm^{-2}$). (b) The nonlinear response of the output power to the peak pump intensity. The relationship between mode spacing D$\lambda$ and nanowire length L in (c) indicates a high group index of 11 due to the high material gain. The pictures on the left correspond to microscope images of a plasmon laser with d = 66 nm, exhibiting spontaneous emission, amplified spontaneous emission, and laser oscillation, where the scattered light output is from the end facets. (d) Threshold intensity of plasmonic and photonic lasers versus the nanowire diameter. The experimental data points correspond to the onset of amplified spontaneous emission, which occurs at slightly lower peak pump intensities compared to the threshold of gain saturation. Amplified spontaneous emission in hybrid plasmonic modes occurs at moderate pump intensities of 10–60 MW $cm^{-2}$ across a broad range of diameters. The authors found a strong increase of spontaneous emission rate, which was associated with Purcell effect of the strong confinement of the plasmon mode. A Purcell factor larger than 5 was achieved with an insulator layer of 5 nm thickness. The authors also estimated the saturated internal gain of the laser from the measurements on averaged group index, and the gain coefficient is over 10,000 $cm^{-1}$, enough to compensate for the propagation loss of the mode, which is around 4100 cm–1, and transmission loss at the end facets of the nanowires.

The same group demonstrated a plasmonic laser operating at room temperature with a subwavelength mode volume. This laser has the same three components in a vertical direction as the one. However, it adopted a squared top layer, which made it benefit from a much-decreased cavity loss, as shown in Figure 8. When the SPP mode is hybridized with the transverse magnetic (TM) mode of CdS square, the effective refractive index increases dramatically, resulting in total internal reflection as it reflects off the edge of the square top. For the photonic mode, on the other hand, its profile becomes increasingly delocalized as the thickness of $MgF_2$ decreases and cannot undergo total internal reflection. Therefore, for a CdS layer of 45 nm and a $MgF_2$ layer of 5 nm, only the plasmonic mode starts to lase as the pump power increases.

The authors also went on showing that the number of lasing cavity modes can be decreased to one by tailoring the shape of the top layer. In this experiment, the Purcell factor is estimated to be 17, with the combination effect of strong mode confinement and improved cavity. The SPP loss is estimated to be 6323 $cm^{-1}$, much smaller than the gain that can be provided by the semiconductor material, which is around 10,000 $cm^{-1}$.

Since the publication of the two papers, hybrid plasmon modes at the semiconductor–insulator–metal interfaces have become a popular model system to achieve SPASER with a well-confined mode volume. Exploiting a triangular cross section of the GaN nanowire, Zhang achieved a hybrid plasmonic laser at an ultraviolet wavelength (370 nm) and with a low lasing threshold at room temperature. Adapted epitaxial-grown Ag as the substrate in an effort to further decrease the lasing threshold of SPASER. They demonstrated a low-threshold hybrid plasmonic laser pumped by a continuous-wave (CW) laser at 120 K. The same authors reported ultralow-threshold CW hybrid plasmonic lasers operating at three primary colors, with a lasing threshold as low as 10 W/cm$^2$.

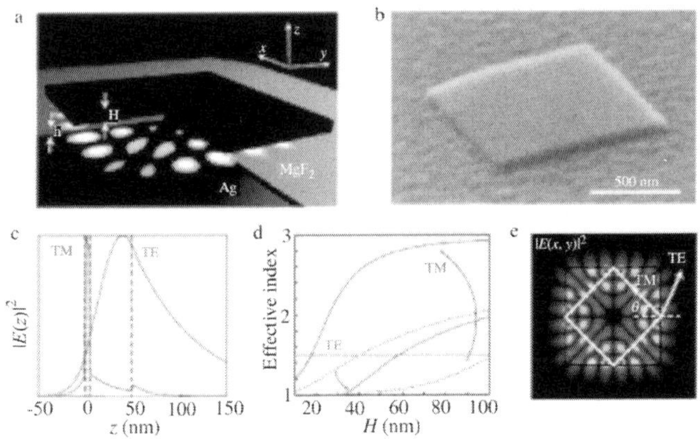


**Figure 8.** The room-temperature plasmon laser.

(a) Schematic diagram of the room-temperature plasmon laser showing a thin CdS square atop a silver substrate separated by a 5 nm $MgF_2$ gap, where the most intense electric fields of the device reside. (b) Scanning electron micrograph of the 45 nm thick, 1 μm length CdS square plasmon laser studied here. (c) The electric field intensity distribution of the two modes of the system along the z direction. Although TM modes are localized in the gap layer, transverse electric (TE) modes are delocalized from the metal surface. (d) The effective index of TM and TE waves with (solid line) and without (dashed line) the metal substrate. TM waves strongly hybridize with SPPs, resulting in strong confinement within the gap region (TM line in (c)) accompanied by a dramatic increase in momentum (solid TM line in (d)) with respect to TM waves of the CdS square alone (dashed TM line in (d)). However, the delocalized TE waves (TE line in (c)) show decreased momentum (solid TE line in (d)) with respect to TE waves of the CdS square alone (dashed TE line in (d)). (e) Electric field intensity distribution of a TM mode in the x and y directions. Although both mode polarizations are free to propagate in the plane, only TM modes have a sufficiently large mode index to undergo efficient total internal reflection, providing the feedback for lasing.

To demonstrate the feasibility of large-scale integrated plasmonic lasers at room temperature, Liu used industry relevant heterogeneous quaternary AlGaInP as the gain material and top-down patterning and etching techniques to demonstrate room-temperature, low-threshold, red-emitting hybrid plasmonic lasers of various cavity shapes, including waveguides, rings, circular disks, and rectangles of a mode volume down to $0.2(l/2n_{eff})^3$ with footprints as small as 350 nm × 390 nm. The authors observed a strongly enhanced spontaneous emission rate with a Purcell factor as large as 29. In particular, the authors attributed the enhanced stimulated cross section observed in the hybrid plasmonic lasers to the mode confinement (Purcell factor). This connection is critical for achieving low lasing threshold at extremely small hybrid plasmonic cavities.

An interesting twist to the vertical layout of semiconductor insulator-metal configuration was reported. The authors coupled a CdSe nanowire with a silver nanowire through near-field contact, so the entire structure forms an X shape. Different from the hybrid plasmonic modes, where the photonic and plasmonic components are hybridized in the transverse direction, here the photon cavity mode (within a CdSe nanowire) and the plasmon cavity mode (within a Ag nanowire) are hybridized along the longitudinal direction and thus spatially separated, as shown in Figure 9. By pumping on the CdSe nanowire portion of the hybrid cavity, the authors can get lasing output from the end facet of the Ag nanowire, with a mode area of 0.008 $\lambda^2$.

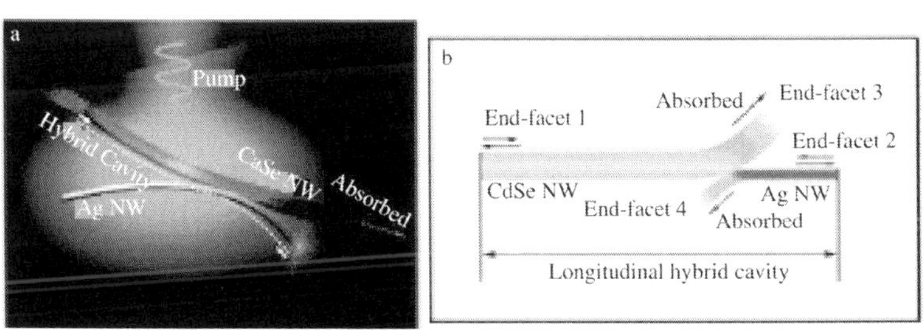


**Figure 9.** Schematic illustration of a hybrid photon-plasmon nanowire

laser. (a) Closed-up view of the coupling area indicating the coupled hybrid cavity (marked by the dashed line), which serves as the hybrid photon-plasmon lasing cavity. (b) Cavity formation in the X structure.

Demonstrated an SPP microcylinder cavity laser. The microcylinder consists of Au or Ag microwires (MWs), whose diameters are 10 and 27.2 μm, respectively. To make the lasers, the authors dipped the MWs into polymethyl methacrylate (PMMA) and R6G solution and then pulled them out and let them dry in air. This process eventually leaves a dried dye-doped PMMA film outside the MWs, and the quality of the film depends on the thickness of the PMMA layer. As a reference, the authors also fabricated the dye-doped PMMA layer outside a fiber glass. The authors observed lasing of multiple whispering gallery modes from the metal MWs up on an optical pump and concluded that these modes are from the propagating SPPs at the inner interface of metal and PMMA film from the spectral spacing of the lasing peaks.

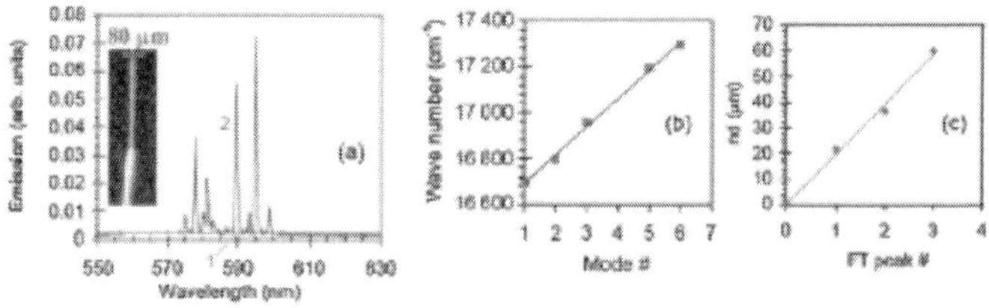


**Figure 10.** (a) Emission spectrum of R6G/PMMA-coated gold wire; trace 1 is with 0.026 mJ and trace 2 is with 0.052 mJ. Inset: microphotograph of the wire; (top part) 10 μm bare gold wire and (bottom part) ~30 μm wire coated with R6G/PMMA. (b) Dependence of the mode energy versus the mode number. (c) Position of the Fourier transform (FT) peak (in units of nd) versus the FT peak number.

### 7.3.2. Amplification of Hybrid Surface Plasmon Polaritons

Despite the tremendous successes of hybrid plasmonic lasers, they can only provide nanoscale light sources. A successful signal transport within the plasmonic circuits requires in situ amplification mechanism to compensate for the signal attenuation due to propagation loss, so the information carried by the plasmonic signal can be transferred from one part of the circuit to another. This is crucial to the scalability and cascadability of the plasmonic circuits.

We experimentally demonstrated in 2013 that a weak hybrid plasmonic signal can be amplified and fully loss compensated by the optical pump-probe technique. The setup to execute the signal amplification through a pump probe is demonstrated in Figure 11. In the experiments, we measured the loss and gain coefficients of a hybrid plasmonic waveguide, consisting of a CdSe nanobelt (NB) separated from a silver surface by an $Al_2O_3$ thin layer, using the pump-probe technique. We demonstrated that these hybrid plasmonic waveguides support an ultrahigh optical gain, with a gain coefficient exceeding 6755 $cm^{-1}$.

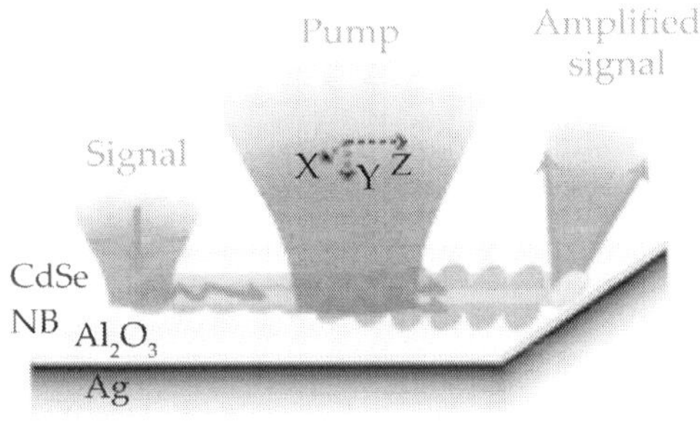


**Figure 11.** Diagram of CdSe NB/$Al_2O_3$/Ag hybrid plasmonic waveguide and the excitation and amplification of the input probe signal when operated in a pump-probe setup.

To characterize how the pump-probe technique can adapt to different polarizations of the probe signal, we alter the polarization of the input signal. This in turn changes the polarization of the waveguided hybrid SP modes. By measuring the output intensity with and without the pump light, we can directly extract the gain coefficient of each hybrid SP mode. Figure 12 shows the signal amplification through stimulated emission on probe signals of orthogonal polarizations. It is quite clear that the (attenuated) probe signals of both polarizations are significantly amplified when the pump light is on. Taking into account the propagation loss of a waveguided signal, we concluded that 99% of the propagation loss of a TM-dominant (input beam polarized along the long axis of the NB) hybrid plasmonic mode is compensated, while only 58% of that of the higher-order mode (input beam polarized perpendicular to the long axis of the NB) is compensated due to the stronger loss in the higher-order mode.

More importantly, we show that the loss compensation works in a relatively broad spectral band and for input signals of tunable polarizations, in contrast to nanolasing that requires single-mode operation. Our result is an important step forward to the realization of integrated plasmonic circuits.

From the experiments discussed in this section, we can clearly that by carefully engineering the gain materials at the close vicinity of metal films or metal wires, the propagation loss of shortrange SPPs caused by the damping of metal can be fully compensated by the gain media, evident by the lasing of SPASER under optical pumping. On the basis of the same principles, amplifiers that can be used to magnify a weak short-range SPP signal with the similar gain coefficients as that for the SPASER have also been demonstrated experimentally. These plasmonic lasers and amplifiers can greatly advance the development of on-chip integrated plasmonic optics.

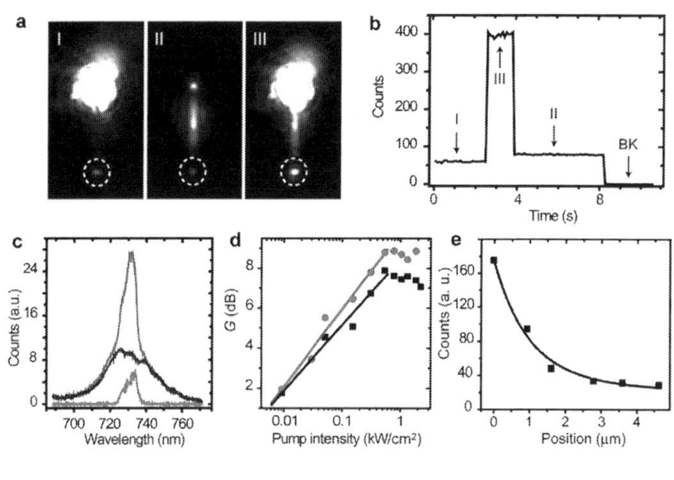


**Figure 12.** Amplification of Hybrid Surface Plasmon Polaritons.

(a) Optical images obtained with a 730 ± 5 nm band pass filter, corresponding to a probe signal launched from the top end of the hybrid plasmonic waveguide (NB 177 nm × 140 nm × 8.6 μm) and emitted from the bottom end, highlighted by the dashed circle (I), photoluminescence (PL) with the pump only (II), and the amplification of the probe signal when both pump and probe are present (III). The two ends of the waveguide are clearly visible as the top and bottom bright spots in (II). (b) Time trace of the output intensity at the emission end of the hybrid Plasmonic waveguide. The dashed circles in (a) mark the area where the time trace was obtained. BK indicates the background dark counts. (c) Typical set of emission spectra, representing the probe signal (bottom), PL (middle), and total output (top). The pump intensity is 0.77 kW/cm2. (d) Gain G (dB) versus pump intensity. The symbols ◎ and ∑ correspond to the measured gains of the input laser beam polarized along and perpendicular to the waveguide, respectively, and the solid lines correspond to logarithmic fits to the G value at a pump intensity lower than 0.6 kW/cm2. (e) Propagation loss measurement. Output light intensity as a function of pump position and fit to linear combination of two exponential decay functions.

### 7.3.3. Surface plasmon polaritons

With the rapid development of optical techniques, data transmission systems require highly increasing integration of photonic devices. It is estimated that the data transmission rates will reach 10 Tbit $s^{-1}$ by decreasing the element size of the photonic devices to the nanometre scale; in this case, the data storage density will be as high as 1 Tbit $inch^{-2}$. However, it is very difficult for conventional photonic devices to reduce the element sizes to the nanometre scale because of Abbe's diffraction limit to about one-half of the optical wavelength. Excitation of surface plasmon polaritons (SPPs) can overcome the diffraction limit and offer a promising approach to control and manipulation propagation and dispersion of light on the nanometre scale. SPP is an electromagnetic excitation existing at the interface between a metal and a dielectric material; furthermore, the resonant interaction between the SPP and the electromagnetic radiation at metallic interfaces results in a remarkably enhanced optical near-field.

SPPs were first observed by Wood in 1902; he found unexplained features in optical reflection measurements on metallic gratings. Then, Maxwell Garnett observed bright colours in metal-doped glasses, and Mie proposed the theory of light scattering by spherical particles in 1908. Pines theoretically described the characteristic energy losses experienced by fast electrons travelling through metals in 1956, and he attributed these losses to collective oscillations of free electrons in the metal, and called these oscillations 'plasmons'. In the same year, Fano introduced the term 'polariton' for the coupled oscillation of bound electrons and light inside transparent media. Thereafter, Ritchie investigated electron energy losses in thin films and performed the first theoretical description of surface plasmons. It was found that plasmon modes could exist near the surface of metals; then, he described the anomalous behaviour of metal gratings in terms of surface plasmon resonances excited on the gratings. Otto and Kretschmann proposed optical excitation of surface plasmons on metal films. Major progress was made when the surface plasmon properties of gold and silver nanoparticles were described by Kreibig and Zacharias for the first time. Several years, Cunningham and co-workers introduced the term surface plasmon polariton (SPP).

In recent years, fundamental and development of SPP-based structures and devices has received increasing interest due to its peculiar properties and applications in optics, surface-enhanced Raman spectroscopy (SERS), data storage, solar cells, sensors, etc. For instance, a new scanning plasmon near-field microscope with a high lateral resolution of 3 nm ($\lambda$/200) was achieved based on the enhanced near-field between sample and metal tips owing to the resonantly excited extended SPP waves. On the other hand, the magnitude of the near-field enhancement was expected to be 100–200 times as large as the incident electromagnetic radiation and indicated a high detection efficiency in near-field spectroscopy.

Five-dimensional (e.g. integration of wavelength, polarization and spatial dimensions) optical recording has been performed by exploiting the unique properties of the longitudinal SPP resonance of gold nanorods. Concretely, the recordings were detected using longitudinal SPP resonance mediated twophoton luminescence, and as a result, a recording disk capacity of 7.2 Tbit and a recording speed of 1 $Gbits^{-1}$ were obtained. SPP-based silver nanoparticles of diameter 40–100 nm can work as a biosensing label, and the detection sensitivity in a DNA hybridization assay is about 60 greater than that achieved using fluorescent labels. Raman signals of adsorbed molecules can be amplified, and using SPP-based silver nanoparticles, the intrinsic Raman enhancement factors are on the order of $10^{14}$ for single Rhodamine 6G molecules adsorbed on selected nanoparticles, which is much larger than the ensemble-averaged values derived from conventional measurements.

## *Surface plasmons in nano-optics*

Scanning near-field optical microscopy as well as fluorescence studies lead to new ways of exciting SPP. The parallel components of the wave vector ($k_x$) necessary for SPP excitation are also present in sub wavelength confined optical near fields in the vicinity of sub wavelength apertures, metallic particles or even fluorescent molecules. If such confined fields are brought close enough to a suitable interface, coupling to SPP can be accomplished locally. Figure 13 shows the principal arrangements. A metal film resides on a (hemispherical) glass prism to allow light (e.g. due to radiation damping of the SPP) to escape and to be recorded. In order to excite surface plasmons, the exciting light field needs to have evanescent field components that match the parallel wave vector $k_x$ of the surface plasmon. As an illustration, Figure 14(a) shows the excitation of surface plasmons with an oscillating dipole placed near the surface of a thin silver film deposited on a glass surface. The figure depicts contourlines of constant power density evaluated at a certain instant of time and displayed on a logarithmic scale. The surface plasmons propagating on the top surface decay radiatively as seen by the wave fronts in the lower medium. The situation is reciprocal to the situation of the Kretschmann configuration discussed earlier where such radiation is used to excite surface plasmons. Also seen in Figure 14 (a) is the excitation of surface plasmons at the metal-glass interface. However, at the wavelength of $\lambda = 370$nm, these plasmons are strongly damped and therefore do not propagate long distance. Figure 14 (b) shows the radiation pattern evaluated in the lower medium (glass). It corresponds to the radiation collected with a high numerical aperture lens and then projected on a photographic plate. The circle in the center indicates the critical angle of total internal reflection of an air-glass interface $\theta_c = \arcsin(1/n)$, with n being the index of refraction of glass. Obviously, the plasmon radiates into an angle beyond $\theta_c$. In fact, the emission angle corresponds to the Kretschmann angle discussed previously. Surface plasmons can only be excited with p-polarized field components as there needs to be a driving force on the free charges towards the interface. This is the reason why the radiation pattern shows up as two lobes.

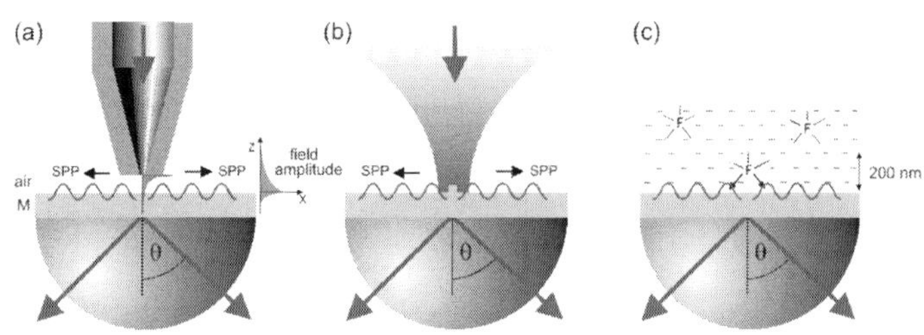


**Figure 13.** Local excitation of surface plasmons on a metal film with different confined light fields. (a) a subwavelength light source such as an aperture probe, (b) an irradiated nanoparticle, and (c) fluorescent molecules. In all cases, surface plasmons are excited by evanescent field components that match the parallel wavevector $k_x$ of the surface plasmon.

The dipole is an ideal excitation source and more realistic sources used in practice have finite dimensions. The size of the source and its proximity to the metal surface determines the spatial spectrum that is available for the excitation of surface plasmons. If the source is too far from the metal surface only plane wave components of the angular spectrum reach the metal surface and hence coupling to surface plasmons is inhibited. Figure 15(a) shows a sketch of the spatial spectrum (spatial Fourier transform) of a confined light source evaluated in planes at different distances from the source. The spectrum is broad close to the source but narrows with increasing distance from the source. The same figure also shows the spatial spectrum of a surface plasmon supported by a silver film. The excitation of the surface plasmon is possible because of the overlap of the spatial spectrum of source and surface plasmon. Due to the decrease in field confinement for increasing distance from the source, a characteristic distance dependence for the surface plasmon excitation efficiency is expected. As discussed before, in a thin film configuration, surface plasmon excitation can be monitored by observing the plasmon's leakage radiation into the glass half space. Figure 15(b) shows, for a thin gold and silver film deposited on a glass hemisphere, the total intensity of surface plasmon leakage radiation as a function of the distance between source (aperture) and the metal surface. The curve labeled 'MMP' indicates a numerical simulation. All curves clearly show a dip for very small distances. This dip is likely due to the perturbation of the surface plasmon resonance condition by the proximity of the probe, i.e. the coupling between probe and sample. For a source with vanishing dimensions (a dipole) no such dip is observed.

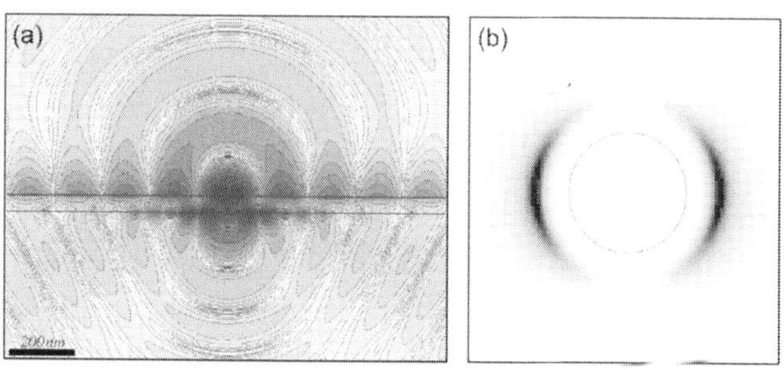


**Figure 14.** Excitation of surface plasmons with a dipole source placed 5nm a 50nm silver layer supported by a glass substrate.

The excitation wavelength is $\lambda$ = 370nm and the dipole moment is parallel to the interface. (a) Lines of constant power density (factor of 2 between successive contour lines) depicted at a certain instant of time. The figure shows the surface plasmon propagation along the top surface of the silver film and also the radiative decay into the lower half space. (b) Radiation pattern evaluated on a horizontal plane in the lower medium. The circle indicates the critical angle of total internal reflection at an air-glass interface. The two lobes result from the radiative decay of surface plasmons excited by the dipole source.

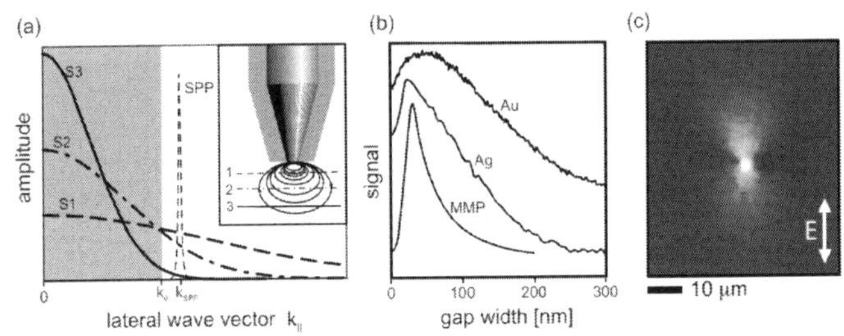


**Figure 15.** Local excitation of surface plasmons with a near-field aperture probe.

(a) Sketch of the overlap of the spatial spectra of source (evaluated in planes at different distances from the source) and the surface plasmon on a silver film. (b) Distance dependence of the coupling. The dip at short distances is a result of probe-sample coupling, i.e. the presence of the probe locally modifies the plasmon resonance condition. (c) Image of plasmon propagation recorded by focusing the leakage radiation on an image plane.

## 7.4. SURFACE PLASMON LASERS: SOURCES OF NANOSCOPIC LIGHT

Metals provide strong optical localization by storing energy on surface electrons that collectively move in unison with light's electric field; excitations known as surface plasmons (SPs). There are two caveats: firstly, practical efforts to focus light far beyond the diffraction limit have struggled to meet theoretical predictions in terms of both efficiency and intensity enhancement; and secondly, having overcome the first challenge, light on metal surfaces is short lived due to electron scattering. The recent demonstration of SP lasers now provides a solution to both challenges. Furthermore, the subtle distinction between SP and photon amplification defines a new exciting class of laser. Plasmonic lasers directly generate nanoscale light, essentially using electronic transitions as a focusing mechanism while also coherently sustaining it by amplification.

### 7.4.1. Surface plasmon lasers

The coupling of energy directly into SPs by emission for fast and efficient light generation at the nanoscale; however, their concept was not implemented experimentally until 2009. The construction of a plasmonic laser is similar to that of a conventional laser. SPs are directly generated on a metal nanostructure and amplified by an adjacent dielectric medium incorporating gain, while a feedback mechanism allows SPs to resonate. The gain medium amplifies SPs by stimulated emission of radiation (SPASER) generating coherently amplified light that is bound to the metal dielectric interface and typically cannot escape without coupling optics. The involvement of electrons in the SP adds momentum to light, confining it tightly to the metal. Plasmonic lasers exploit this confinement effect to deliver intense optical energy well the diffraction barrier on extremely fast time scales, thus extending the capabilities of laser systems in general.

The first plasmonic lasers utilized quite different metallic nanostructures. Hill and co-workers used a metalized, etched semiconductor heterostructure, as shown in Figure 16a. Surface plasmons between the two metal planes were also weakly confined by a small index variation along the heterostructure's growth direction leading to guided wave propagation parallel to the plane of the substrate. The thinnest heterostructure reported in this work was just 90 nm thick leading to a SP smaller than the diffraction limit in one dimension, i.e. between the metal planes. The end facets of the metalized structure completed a Fabry-Perot SP resonator. In their experiments, a small amount of scattering from the un-metalized base allows far-field detection and characterization of the laser through the substrate.

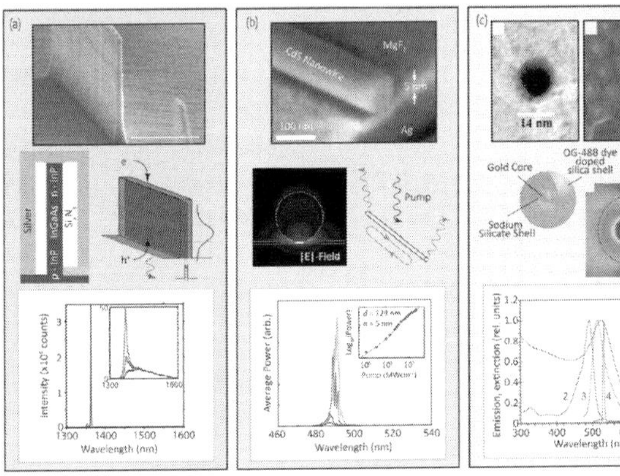


**Figure 16.** Scanning electron micrographs and emission spectra of plasmonic lasers exhibiting sub-diffraction limited confinement in (a) one, (b) two and (c) all three dimensions. Each panel shows a scanning electron image of the device, a schematic indicating the shape of the laser mode's electric field intensity and emission spectra below and above threshold.

Another device reported by Oulton and co-workers also used a Fabry-Perot cavity, but with plasmonic confinement in two dimensions perpendicular to wave propagation. The laser cavity consists of a semiconductor nanowire sitting on a flat metal film with a thin nanoscale insulating gap, as shown in Figure 16b. Remarkably, the mode of the nanowire and the SP of the metal surface hybridizes into a deep sub-wavelength mode propagating along the wire's axis. Here, a small amount of feedback was provided by the end facets of the nanowire forming a cavity. Similar to Hill's device, this one generates optical energy principally bound to the metal plane; however, the open geometry allows light extraction to the far-field due to scattering from the nanowire end-facets.

These plasmonic lasers resemble conventional lasers in that they use a Fabry-Perot cavity to generate feedback; however, nano-optics can even circumvent this tenet of laser physics.

Since metallic particles just a few nanometers in size support localized SPs, in principle, plasmonic lasers could be as small. Indeed, Noginov and co-workers have reported a nano-particle laser just 44 nm in diameter. Here a 15 nm diameter Gold nano-particle supports a dipole like plasmonic "cavity" mode while a Silica shell impregnated with OG-488 dye molecules serves as the gain medium, as shown in Figure 16c. Since the particle is considerably smaller than the wavelength, radiation loss is small and the localized SP loss is almost entirely due to absorption. The vast majority of the generated light is stored at the metal-dielectric interface. The authors report the signature of laser action under strong optical excitation. However, out of all of the plasmonic lasers reported to date, this is the only configuration that has yet to be demonstrated again. Since there is a growing interest in the use of gain with arrays of such nano-particles towards loss-less metamaterials, more into the laser action of metal particles is extremely important.

These demonstrations show clearly the principal functionality of plasmonic lasers: they deliver energy to the nanoscale, releasing only a small part of their optical energy to far field radiation before it dissipates in the metal. Furthermore, in stark contrast to the highly directional beams of conventional lasers, the small amount of radiation coupled to the far-field strongly diffracts. These characteristics highlight the challenge of measuring plasmonic laser emission and their unique use as sources of nanoscopic light.

## 7.4.2. The Anatomy of a Plasmonic Laser

### *Surface plasmon waveguides*

There are a variety of metallic nano-structured waveguides capable of strong confinement, detailed in a recent comprehensive review. While many show quite subtle physics, generally optical confinement relies on two mechanisms. The first occurs near the SP frequency, $\omega_{sp}$, where the coupling between electrons in the metal and light in the adjacent dielectric is strongest; even a flat metal-dielectric interface is capable of strong confinement. A Drude model of a metal's permittivity, $\varepsilon_m(\omega)=\varepsilon_b-\omega_p^2/\omega(\omega-i\gamma)$, gives $\omega_{sp}=\omega_p/\sqrt{\varepsilon_b+\varepsilon_d}$ at the peak in optical momentum of SPs of a flat metal-dielectric interface. $\omega_{sp}$ is related to the metal's plasma frequency, $\omega_p$, and the background permittivity's of the metal, $\varepsilon_b$, and the adjacent dielectric, $\varepsilon_d$. The additional electron momentum gives SPs much shorter wavelengths (higher momentum) than is possible with purely dielectric materials. It is the shorter SP wavelength that also generates strong lateral confinement. The SP frequency, the coupling strength between light and electrons diminishes leading to surface waves that travel almost as fast as waves in the dielectric. This is the origin of the characteristic dispersion curve of SPs, shown in Figure 17c. Unfortunately, lateral confinement also diminishes with the momentum at lower frequencies leaving SPs with spatial extents larger than the vacuum wavelength, even in the near infrared.

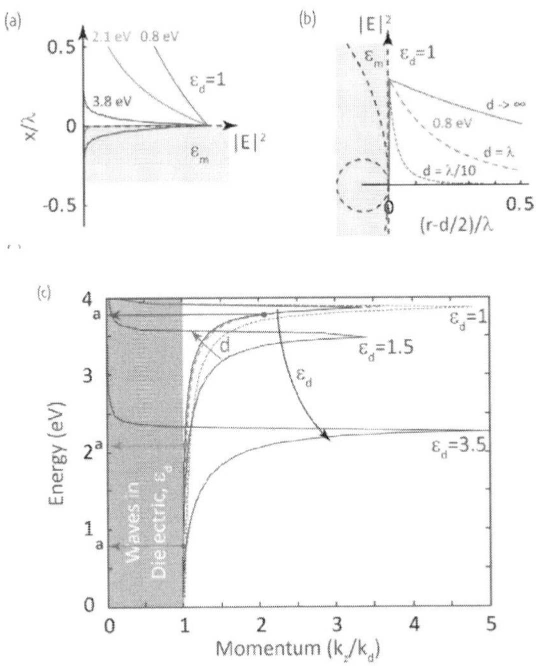

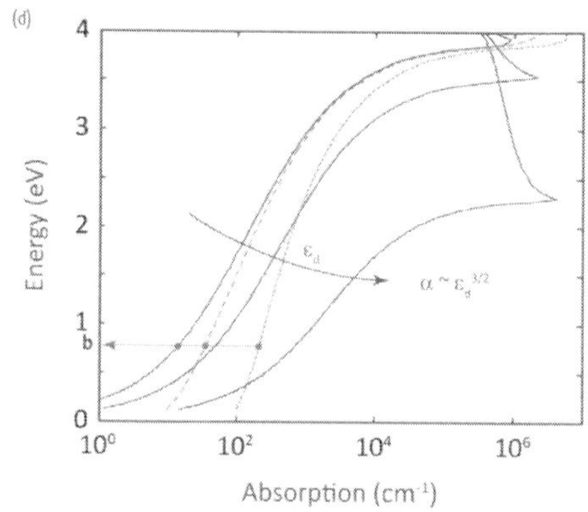


**Figure 17.** Anatomy of surface plasmons.

Distribution of electric field intensity near (a) a flat silver surface and (b) the curved surface of a metal cylinder adjacent to air. The dispersion relations for these structures and for various dielectric permittivities, $\varepsilon_d$, are shown in (c). The loss incurred by electron scattering in the metal is shown in (d) showing the increase in loss with both degree of confinement and $\varepsilon_d$.

Confinement at lower frequencies requires the second mechanism. All SPs generate a strong surface charge at metal-dielectric interfaces as they propagate. Strong confinement can occur when two nearby metaldielectric interfaces are capacitively coupled through this charging; e.g., between the two metal planes, like in Hill's plasmonic laser. Even a small amount of curvature on a metal's surface can restore confinement; e.g., at the apex of metal wedges or grooves or for cylindrical metal wires. Figure 17b shows how rapidly the SP field distribution shrinks with a metal plane's decreasing radius of curvature. Such capacitive optical energy storage is possible anywhere the surface plasmon frequency, even at the frequency of mains electricity. As a rule of thumb, a structure that exhibits a strong electrostatic capacitance should also provide good confinement of propagating SPs. Meanwhile, there is only a modest increase of the SP momentum with increasing confinement, in contrast to SPs near $\omega_{sp}$ (Figure 17c). This clearly distinguishes the roles of electron-light coupling near $\omega_{sp}$ and capacitively coupled surface charges on confinement.

While the first mechanism relies on electrons as deep as the metal's skin depth to increase light's momentum, the second is purely a surface effect where only electrons within the Thomas-Fermi screening distance of the interface contribute to surface charge.

The principal problem with the use of metals to achieve confinement is high loss. Figure 17d shows the loss described by the Drude model due to electron scattering. Clearly, loss near the SP frequency is strongest suggesting that confinement based on capacitive coupling is preferable. However, capacitive confinement does not escape the problem of loss; although the cylinder's curvature makes only a modest impact on the SP momentum, the stronger confinement at lower frequencies incurs more loss as proportionally more light propagates within the metal.

The absorption coefficient of SPs also depends on the adjacent dielectric material. Figure 17d shows the losses of SPs of a flat metal interface with air ($\varepsilon_d = 1$), insulator ($\varepsilon_d = 1.5$) and semiconductor ($\varepsilon_d = 3.5$). While losses are relatively insensitive to $\varepsilon_d$ near the $\omega_{sp}$, at low frequencies the absorption increases with at least $\varepsilon_d$ 3/2; e.g., SPs of air-metal interfaces propagate more than forty times further than those on semiconductor-metal interfaces. Since semiconductors provide well established optoelectronic functionality, in particular the strong gain necessary to realize plasmonic lasers, overcoming this critical flaw in metal-based optics is crucial.

### *Loss management of semiconductor-based plasmonics*

Surface plasmon loss becomes a significant problem when both nano-structuring and high permittivity materials are used.

The problem is best illustrated by a parametric plot of propagation distance versus confinement (Figure 18). Consider three waveguide geometries at a wavelength of 1500 nm: a dielectric clad metal cylinder; a dielectric hole in metal (parameterized by their diameters, d); and a hybrid of a semiconductor cylinder separated from a metal plane by an insulating gap (parameterized by the gap thickness, h, for a given cylinder diameter, d). In the first two cases, the dielectric cladding is either ($\varepsilon_d$ = 1.5) or semiconductor

($\varepsilon_d = 3.5$). The shaded region in Figure 18 highlights where modes are larger than the diffraction limited area. The black arrows on the left hand side show the propagation lengths of SPs at Insulator/Silver and Semiconductor/Silver planar interfaces. The black arrow on the top axis shows the mode area for an uncoupled insulator clad semiconductor cylinder.

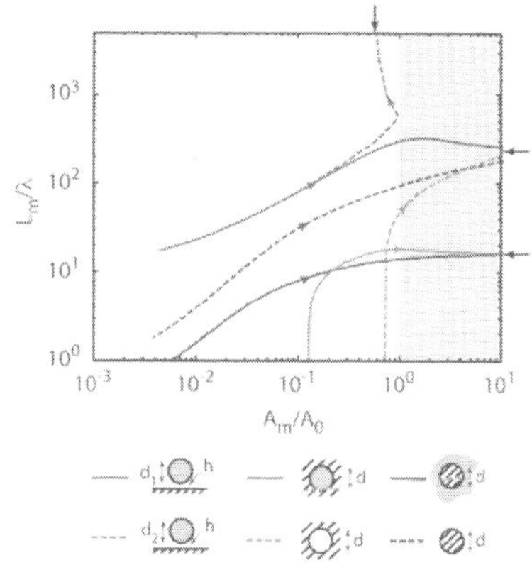


**Figure 18.** Parametric plot of normalized mode area and normalized propagation distance for a hybrid waveguide (red lines), oxide and semiconductor clad silver cylinders (blue lines) and holes of insulator and semiconductor in silver (green lines). Each trajectory represents a waveguide for a range of structural parameters. For the hybrid waveguide, $h = 1 \rightarrow 10^4$ nm for $d_1 = 200$ nm and $d_1 = 215$. (The two diameters highlight the two possible asymptotic solutions as $h \rightarrow \infty$: pure SPP and cylinder modes.) For the other waveguides, the diameters, $d = 10^1 \rightarrow 10^4$ nm are used.

The metal cylinder supports a mode with deep sub-wavelength characteristics for a range of diameters. (The dielectric hole in metal is diffraction limited and has a mode cut-off.) The price of a mode with 1/10th of the diffraction limited area is an order of magnitude shorter propagation distance than a flat insulator-metal interface. With semiconductor cladding, there is very little benefit in terms of increased confinement and the propagation distances decreases by almost another order of magnitude. Such a waveguide can barely sustain SPs for practical purposes even with a near diffraction limited mode area.

Currently, the only solution to the problem of loss in metal nanostructures incorporating semiconductors is to carefully control the coupling of light and electrons by mode hybridization. For example, Figure 18 shows the trajectory of a mode formed by the coupling of a semiconductor nanowire's fundamental mode with the SP of a flat metal plane. By ensuring that the semiconductor does not come into contact with the metal, the propagation loss can be drastically reduced; even at maximal confinement for small gap widths, the propagation loss does not exceed that of SPs of a flat semiconductor-metal interface. Furthermore, this structure provides additional confinement; just as metal-dielectric interfaces charge with electrons as a mode propagates, the interfaces of high and low permittivity materials can polarize. The interaction of surface charge on the metal and the surface polarization on the semiconductor generates strong capacitive localization in the insulating gap region. In this case, this hybrid plasmonic mode attains similar confinement to an insulator clad metal nanowire, but propagates almost an order of magnitude further. Hybrid plasmonics are a practical approach to truly nanoscale semiconductor-based plasmonics.

### *Surface plasmon feedback*

In order to construct a plasmonic laser, some form of SP feedback is required. For plasmonic systems that utilize mainly capacitive light-electron coupling, a conventional cavity can be used, such as a Fabry-Perot cavity formed between two mirrorsor a whispering gallery cavity. Typically, no additional confinement beyond the diffraction limit in the third dimension is possible because the SP momentum is generally not much larger than that of light in the dielectric. A number of conventional feedback mechanisms have been employed in plasmonic lasers. Hill and Oulton use a Fabry-Perot cavity approach. More recent work by Hill and co-workers also show the effectiveness of distributed feedback by incorporating a grating in the sidewalls of their metal coated etched heterostructure. A plasmonic crystal cavity laser was also recently reported. Another promising approach is to use total internal surface plasmon reflection; this has been demonstrated in both square and cylindrical laser cavities. These are so promising because they have already reached the diffraction limit with SP modes that exhibit dipole like characteristics. Such dipole lasers are still subject to near wavelength scales in a number dimensions since they operate far from the SP frequency. Nevertheless, these lasers are extremely compact, even in the THz range.

Further reduction in both the physical laser size and its mode size is possible near to the SP frequency. Nearer the SP frequency, the additional electron momentum can lead to substantial reductions in the size of the SP mode in the third dimension, as shown in Figure 19. For example, far from the SP frequency, a cylindrical rod's dipole mode follows a half SP wavelength condition, which depends on radius. However, with the stronger lightelectron coupling at bluer wavelengths, the dipole resonance occurs for much shorter rod lengths than the wavelength would suggest, as shown in Figure 19a and 19b. At this point the electrons de-couple from the radiation field and the corresponding localized SP mode becomes insensitive to the optical phase; a localized SP resonance occurs red-shifted from $\omega_{sp}$. Here, the resonance depends solely on the nanoparticles's geometry and the absorptive loss becomes independent of the surrounding dielectric, as shown in Figure 19c. Evidentially, the resonance frequency is independent of the particle's radius and length. As shown by Noginov and co-workers, lasers constructed from such SP resonances could be reduced to the scale of just tens of nanometers.

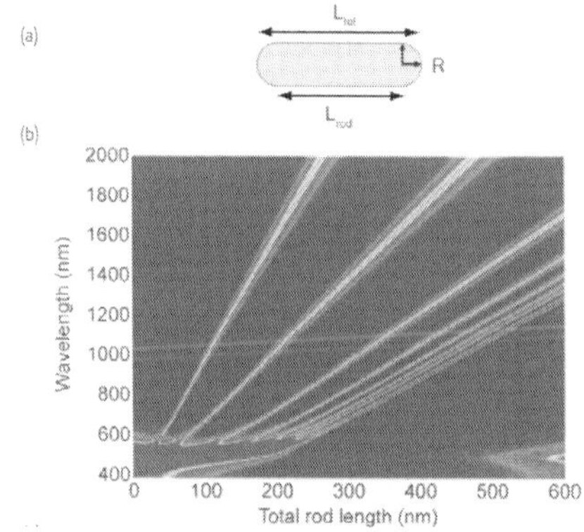

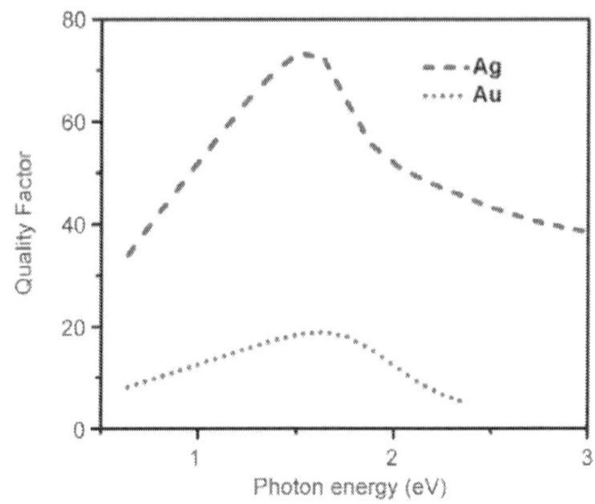


**Figure 19(a).** Schematic and dimensions of a cylindrical nanowire with hemispherical caps supporting a dipolar resonance. (b) The full dependence of the dipole resonance wavelength on Ltot for different R obtained from the condition where Ltot is equal to half a SP wavelength. (c) Quality factor of plasmon resonances of a metal nanostructure as a function of plasmon frequency for two commonly used metals: gold (dashed line) and silver (dotted line).

## *Optical gain media*

Surface plasmon amplification has attracted a great deal of attention as the means to compensate for the loss of SP waveguides, but the choice of suitable gain media is a challenge. Semiconductors provide strong gain, but also introduce additional losses without a hybrid plasmonic approach, as discussed earlier. A number of works therefore sought to amplify relatively weakly confined SPs with gain media based on erbium doped glass, molecules, and quantum dots. These schemes were attractive because they simply required doping any of a variety of low refractive index materials with light emitters; however, gain is typically limited due to low emitter concentrations. At telecoms wavelengths, authors have studied loss compensation of long and short range SPs. Long range SPs are relatively delocalized with loss typically $< 10\ cm^{-1}$ at telecoms wavelengths. While this exceeds the typical gain of erbium doped glass ($\sim 1\ cm^{-1}$), Berini and de Leon were able to observe net amplification using IR140 dye with a maximum material gain of ~ 360 $cm^{-1.}$ Meanwhile, the loss of SPs confined to approximately the diffraction limit is typically many ~ 100 $cm^{-1}$ at telecoms wavelengths. Grandidier and co-workers were unable to compensate dielectric loaded plasmonic waveguide losses of ~ 750 $cm^{-1}$, with gain from quantum dots that only extended up to ~ 200 $cm^{-1}$. At visible frequencies, SP loss is much higher; however, Gather and co-workers were still able to observe net optical gain of long range SPs using a light emitting polymer. Here passive SP loss of ~ 60 $cm^{-1}$ was compensated by a poly-(phenylene-vinylene) derivative with a maximum material gain of ~ 90 $cm^{-1}$. These studies show that low refractive index gain media are certainly capable of compensating long range SP losses of up to a couple of 100 cm-1 in the visible and IR.

Although a number of studies have used molecular gain to compensate for the optical loss of strongly confined SPs, semiconductors tend to be the materials of choice for losses in excess of 1000 $cm^{-1}$. For example, plasmonic nanowire lasers operating at a wavelength of about 500 nm have passive losses of ~ 5000 $cm^{-1}$. In laser experiments, a similar degree of modal gain was observed with an estimated material gain of 10 000 $cm^{-1}$ at threshold due to the partial overlap of gain and mode. In this case the cadmium sulfide semiconductor provided sufficient gain to achieve lasing at room temperature; as an indication, cadmium sulfide has an absorption coefficient in excess of 105 $cm^{-1}$ at 500 nm. At infrared wavelengths, the loss of maximally confined hybrid SPs are on the order of 1000 cm-1, which is well within the gain range of a variety of common compound semiconductors. For example, a number of recently demonstrated metal lasers have used various indium arsenide, gallium arsenide, and indium phosphide alloys for gain near 1.5 μm, built on indium phosphide substrates. While the available semiconductor materials clearly present sufficient gain, the consequences of operating so close to gain saturation is broadly unknown. However, strong gain effects, such as gain guiding, could provide additional design freedoms for future devices.

### 7.4.3. Modified Spontaneous Emission near Metallic Nano-structures

The extremely small optical modes associated with metallic nanostructures drastically modify the rate of spontaneous emission, which in turn modifies the laser process. It is critical to account for this modification as it can influence a laser's threshold by orders of magnitude.

#### *The Purcell effect*

Purcell noted that light-matter interaction rates can be enhanced by reducing the optical mode volume and increasing the Quality factor (Q) of a resonant cavity. Since interaction rates are proportional to the cubed ratio of the respective characteristic lengths scales of optical and electronic wave functions, enhancements of up to nine orders of magnitude are available at visible frequecnies. Purely semiconductor-based photonics has achieved substantial enhancements in spontaneous emission, stimulated Raman scattering, and nonlinear frequency conversion but, having reached near-diffraction-limited sizes, these enhancement rely on high cavity Q. Surface plasmons, on the other hand, exhibit a significant Purcell effect even with a low cavity Q, or even without a cavity at all, by accessing considerably smaller mode sizes: e.g., Figure 20 shows the strong Purcell effect associated with hybrid plasmonic modes. When operating as laser devices, modest Purcell factors up to 6 are found since recombination in the nanowire does not experience the modal electric field maximum. Meanwhile, more recent experiments show Purcell factors as high as 60 for molecules situated within the high intensity gap region between nanowire and metal substrate. Recent has shown Purcell factors of many orders of magnitude at visible frequencies in plasmonic nano-structures.

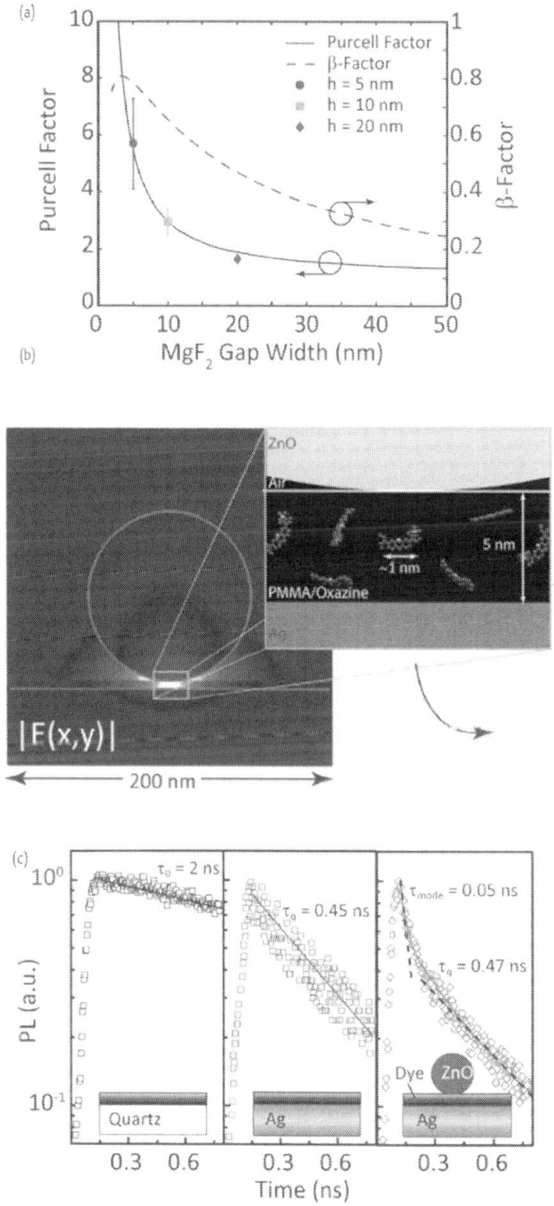


**Figure 20.** The Purcell effect for semiconductor nanowires on Silver.

(a) Purcell enhancement factor for emission from Cadmium Sulfide nanowire I2 excitons at T = 10 K[24]. (b) The field distribution of the strongly localized hybrid plasmonic mode that causes the enhancement at the wavelength of 489 nm, corresponding to $I_2$ exciton recombination at 10K. Since CdS excitons recombine within the nanowire, they do not experience the strongest field in the insulating gap region. The inset shows a schematic of another experiment with molecules placed within the gap region. (c) The emission rate enhancement of a 5 nm thick film of oxazine molecules placed within the strongest field position beneath nanowire. From left to right: oxazine on quartz, oxazine on metal and oxazine between a Zinc Oxide nanowire and metal.

### *Spontaneous Emission Factor*

Metallic nano-structures generally enhancement for just a few of their optical modes. While the most confined modes will exhibit fast emission coupling rates, less well confined modes have relatively slower coupling rates. This leads to preferential coupling to the most confined modes. The spontaneous emission factor, β, is the proportion of emission that couples into a single mode and is extremely important in the threshold behavior of a laser. A simple rate equation analysis shows that a laser's threshold is at least the total cavity loss rate divided by β. For conventional semiconductor lasers, $\beta < 10^{-3}$, while for the smallest metal nanostructures it could be as high 0.8. Consequently conventional lasers ($\beta \ll 1$), require very low cavity losses to ensure realistic thresholds; however, SP lasers can operate with much higher cavity losses provided $\beta \sim 1$.

Metal nanostructures introduce a further complication: emission near a metal surface strongly couples to lossy surface waves (LSWs). Rather than being discrete guided modes, LSWs form a continuum that rapidly dissipates energy into the metal, essentially quenching useful emission. This drastically reduces β and inhibits plasmonic lasers from operating. Figure 21 shows the Purcell effect for plasmonic nanowire lasers as a function of both nanowire diameter, d and insulator gap width, h. Although nanowires in contact with the metal show much larger Purcell enhancements, none of these nanowires achieved laser action. Only nanowires with an insulating gap achieved laser action, even if this gap was only 5 nm thick.

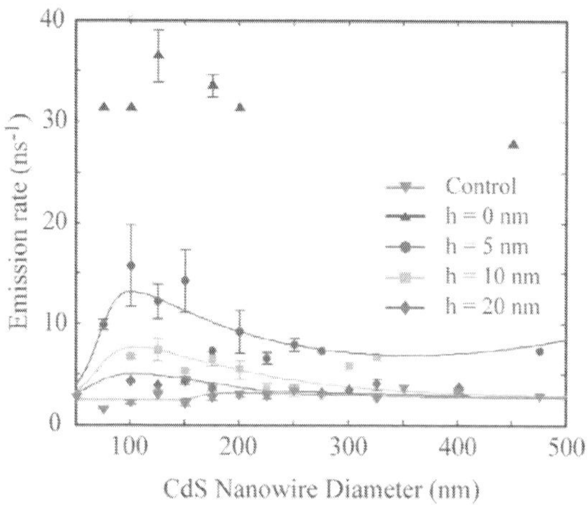


**Figure 21.** Spontaneous emission rates of Cadmium Sulfide plasmonic nanowire lasers as a function of the nanowire diameter and insulator gap width, h. Nanowires with no gap layer had much higher emission rates, but did not lase. Without a gap layer, the quenching of emission near the metal surface by lossy surface waves significantly reduces emission into the laser mode (β factor) making laser threshold unobtainable. Only devices with a gap layer could laze, even if this was just 5 nm thick.

All plasmonic lasers, demonstrated to date, incorporate a thin insulating gap separating the plasmon supporting metal interface and the gain medium. In the case of semiconductor based plasmonic lasers the gap region serves two purposes: it is essential to mitigate both intrinsic plasmonic losses due to electron collisions in the metal and it ensures a high β factor by preventing quenching to LSWs.

### 7.4.4. Recent developments in plasmonic semiconductor lasers

Currently, plasmonic lasers based on inorganic semiconductors are the most technologically relevant being capable of electrical injection and compatible with computing and communications infrastructure. The prospect of compact light sources with footprints comparable to electronics has led to a number of recent works on metal based lasers. Although some devices do not utilize

SPs, these studies significantly improve compact laser technology by addressing issues of electrical contacting and heat sinking, which have been problematic for microscopic semiconductor lasers. Hill and co-workers demonstrated the first nanoscale electrically driven metal coated laser. While the geometry is based loosely on the vertical cavity surface emitting design, the cavity mode actually resonates perpendicular to the surface normal where scattering leads to surface emission. Recent work has seen electrically driven devices, even under continuous wave operation, and optically pumped devices at room temperature, but devices with both qualities have only been possible in larger structures where optical involvement of the metal is greatly diminished. Recently, the first sub-wavelength continuous wave operation device operating at room temperature was reported. Larger devices near the diffraction limit are therefore quite close to commercial viability.

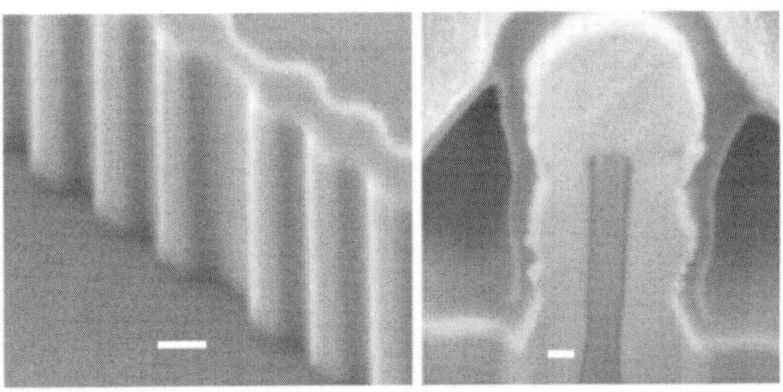

**Figure 22.** Plasmonic distributed feedback lasers have both strong feedback and confinement operating with electrical pumping. While the smallest nanostructures with highly localized plasmonic modes are close to operating electrically at room temperature, larger devices that support non-plasmonic modes are already close to commercial viability. Scale bars represent 100 nm.

There have also been recent works on the amplification and laser action of SPs. Room temperature operation has been reported for long range SPs exhibiting low confinement and for tightly confined hybrid SPs. Meanwhile, the work of Hill and co-workers continues to pioneer electrically injected plasmon lasers. However, the prospect of room temperature operation under electrical pumping with extremely strong optical confinement remains a fundamental challenge.

In addition to the laser action of directly generated SPs, there has been growing interest in adapting plasmonic technology to commercial lasers. For example, metallic nanostructures have been positioned on the facets of commercial quantum cascade lasers to allow strong nanofocusing using antennas and the control of beam directionality using arrays of slits and gratings. These are potentially interesting areas for new developments in the science of plasmonic lasers.

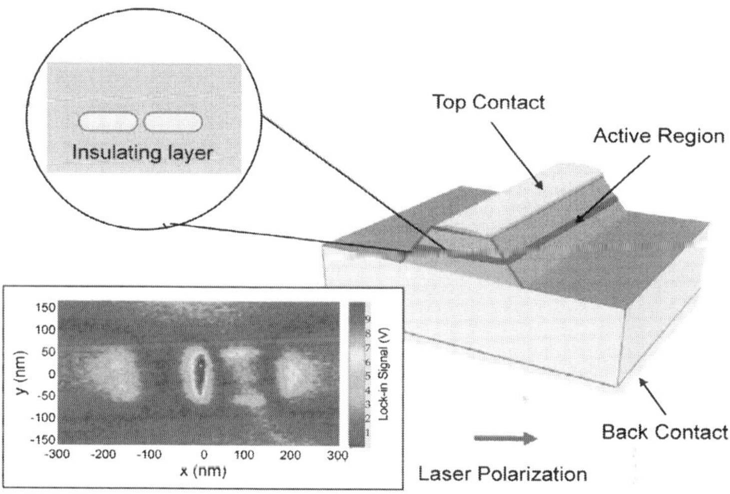


**Figure 23.** An alternative strategy to plasmonic lasers is to add metal nanostructures to commercial laser diodes. Here, a bow-tie nano-antenna on the facet of a quantum cascade laser confines light to nanoscale dimensions.

## REFERENCES

1. A. V. Akimov, A. Mukherjee, C. L. Yu, D. E. Chang, A. S. Zibrov, P. R. Hemmer, H. Park, and M. D. Lukin, Nature, 450, 402–406 (2007).
2. H. Wei, Z. P. Li, X. R. Tian, Z. X. Wang, F. A. Cong, N. Liu, S. P. Zhang, P. Nordlander, N. J. Halas, and H. X. Xu, Nano Lett., 11, 471–475 (2011).
3. H. Wei, Z. X. Wang, X. R. Tian, M. Käll, and H. X. Xu, Nat. Commun., 2, 387 (2011).
4. M. A. Noginov, A. M. Belgrave, R. Bakker, V. M. Shalaev, E. E. Narimanov, S. Stout, E. Herz, T. Sutewong, and U. Wiesner, Nature, 460, 1110–1112 (2009).
5. M. A. Noginov, G. Zhu, M. Bahouza, J. Adegoke, C. Small, C. Darvison, V. P. Drachev, P. Nyga, and V. M. Shalaev, Phys. Rev. B, 74, 184203 (2006).
6. M. C. Gather, K. Meerholz, N. Danz, and K. Leosson, Nat. Photonics, 4, 457–461(2010).
7. N. Liu, A. Gocalinska, J. Justice, F. Gity, I. Povey, B. McCarthy, M. Pemble, E. Pelucchi, H. Wei, C. Silien, H. X. Xu, and B. Corbett, Nano Lett., 16, 7822–7828 (2016).
8. N. Liu, H. Wei, J. Li, Z. X. Wang, X. R. Tian, A. L. Pan, and H. X. Xu, Sci. Rep., 3, 1967 (2013).
9. P. Berini and I. De Leon, Nat. Photonics, 6, 16–24 (2012).
10. Q. Zhang, G. Y. Li, X. F. Liu, F. Qian, Y. Li, T. C. Sum, C. M. Lieber, and Q. H. Xiong, Nat. Commun., 5, 4953 (2014)
11. R. F. Oulton, V. J. Sorger, D. A. Genov, D. F. P. Pile, and X. Zhang, Nat. Photonics, 2, 496–500 (2008).
12. R. F. Oulton, V. J. Sorger, T. Zentgraf, R.-M. Ma, C. Gladden, L. Dai, G. Bartal, and X. Zhang, Nature, 461, 629–632 (2009).
13. R. M. Ma, R. F. Oulton, V. J. Sorger, and X. Zhang, Laser Photonics Rev., 7, 1–21 (2013).
14. S. A. Maier, Plasmonics: Fundamentals and Applications (Springer, New York, 2007).

15. V. M. Shalaev and S. Kawata, Advances in Nano-Optics and NanoPhotonics (Elsevier, Amsterdam, 2007).
16. V. P. Drachev, W. T. Kim, V. P. Safonov, V. A. Podolskiy, N. S. Zakovryashin, E. N. Khaliulin, V. M. Shalaev, and R. L. Armstrong, J. Mod. Opt., 49, 645– 662 (2002).
17. X. Q. Wu, Y. Xiao, C. Meng, X. N. Zhang, S. L. Yu, Y. P. Wang, C. X. Yang, X. Guo, C. Z. Ning, and L. M. Tong, Nano Lett., 13, 5654–5659 (2013).
18. Y. J. Lu, C. Y. Wang, J. Kim, H. Y. Chen, M. Y. Lu, Y. C. Chen, W. H. Chang, L. J. Chen, M. I. Stockman, C. K. Shih, and S. Gwo, Nano Lett., 14, 4381–4388 (2014).
19. Y. J. Lu, J. Kim, H. Y. Chen, C. H. Wu, N. Dabidian, C. E. Sanders, C. Y. Wang, M. Y. Lu, B. H. Li, X. G. Qiu, W. H. Chang, L. J. Chen, G. Shvets, S. Gwo, and C. K. Shih, Science, 337, 450–453 (2012).
20. Y. R. Fang, Z. P. Li, Y. Z. Huang, S. P. Zhang, P. Nordlander, N. J. Halas, and H. X. Xu, Nano Lett., 10, 1950–1954 (2010).

# INDEX

## T



## U



## V



## W